Solutions Manual
to Accompany

MOLECULAR
CELL BIOLOGY

Seventh Edition

Tom Huxford, Stephanie Bingham, Brian Sato,
Steve Amato, Greg Kelly, Tom Keller, Elizabeth Good,
Cynthia Klevickis, Leah Haimo, Brian Storrie,
Eric A. Wong, Richard A. Walker, Glenda Gillaspy,
Jill Sible, and Muriel Lederman

W. H. Freeman and Company
New York

W.H. Freeman and Company
41 Madison Avenue
New York, NY 10010
Houndmills, Basingstoke RG21 6XS, England

www.whfreeman.com

Table of Contents

2

CHEMICAL FOUNDATIONS

REVIEW THE CONCEPTS

1. Less energy is required to form noncovalent bonds than covalent bonds, and the bonds that stick the gecko's feet to the smooth surface need to be formed and broken many times as the animal moves. Since van der Waals interactions are so weak, there must be many points of contact (a large surface area) yielding multiple van der Waals interactions between the septae and the smooth surface.

2. a. These are likely to be hydrophilic amino acids, and in particular, negatively charged amino acids (aspartate and glutamate), which would have an affinity for K^+ via ionic bonds.

   b. Like the phospholipid bilayer itself, this portion of the protein is likely to be amphipathic, with hydrophobic amino acids in contact with the fatty acyl chains and hydrophilic amino acids in contact with the hydrophilic heads.

   c, d. Since both the cytosol and extracellular space are aqueous environments, hydrophilic amino acids would contact these fluids.

3. At pH = 7.0, the net charge is –1 because of the negative charge on the carboxyl residue of glutamate (E). After phosphorylation by a tyrosine kinase, two additional negative charges (because of attachment of phosphate residues to tyrosines (Y)) would be added. Thus, the net charge would be –3. The most likely source of phosphate is ATP since the attachment of inorganic phosphate (P_i) to tyrosine is energetically highly unfavorable, but when coupled to the hydrolysis of the high-energy phosphoanhydride bond of ATP, the overall reaction is energetically favorable.

4. Disulfide bonds are formed between two cysteine residue side chains. The formation of disulfide bonds increases the order and therefore decreases the entropy (S becomes more negative).

5. Stereoisomers are compounds that have the same molecular formula but are mirror images of each other. Many organic molecules can exist as stereoisomers because of two different possible orientations around an asymmetric carbon atom (e.g., amino acids). Because stereoisomers differ in their three-dimensional orientation and because biological molecules interact with one another based on precise molecular complementarity, stereoisomers often react with different molecules, or react differently with the same molecules. Therefore, they may have very distinct physiological effects in the cell.

6. The compound is guanosine triphosphate (GTP). Although the guanine base is found in both DNA and RNA, the sugar is a ribose sugar because of the 2′ hydroxyl group. Therefore, GTP is a component of RNA only. GTP is an important intracellular signaling molecule.

7. At least three properties contribute to this structural diversity. First, monosaccharides can be joined to one another at any of several hydroxyl groups. Second, the C-1 linkage can have either an α or a β configuration. Third, extensive branching of carbohydrate chains is possible.

8. What is the pH of 1 L of water? In all aqueous solutions, water spontaneously dissociates into hydrogen and hydroxide ions according to the equilibrium reaction $H_2O \rightleftharpoons H^+ + OH^-$. The ionization constant for aqueous solutions at 25°C is $K_w = [H^+][OH^-] = 1 \times 10^{-14} \ M^2$. In a solution of pure water, the production of one H^+ ion will always be accompanied by the production of one OH^- ion. In other words, $[H^+] = [OH^-]$.

$$K_w = [H^+][OH^-] = [H^+]^2 = 1 \times 10^{-14} \ M^2$$
$$[H^+] = 1 \times 10^{-7} \ M$$
$$pH = -\log_{10}[H^+] = -\log_{10}(1 \times 10^{-7}) = 7$$

What is the pH after 0.008 moles NaOH are added? NaOH (sodium hydroxide) is a strong base. This means all the added NaOH ionizes to increase the $[OH^-]$ concentration to 0.008 M.

$$[H^+] = K_w/[OH^-] = (1 \times 10^{-14} \ M^2)/(0.008 \ M) = 1.25 \times 10^{-12} \ M$$
$$pH = -\log_{10}[H^+] = -\log_{10}(1.25 \times 10^{-12}) = 11.903$$

What is the pH of the solution of 50 mM MOPS? MOPS is a weak acid. As such, upon dissolving in water it will undergo partial dissociation yielding equal concentrations of hydrogen ions and MOPS conjugate base according to the equilibrium reaction:

MOPS(weak acid form) = H^+ + MOPS(conjugate base form)

The extent to which this reaction goes forward determines the relative strength of the MOPS weak acid and is given by its acid dissociation equilibrium constant:

$$K_a = ([H^+][MOPS(\text{conjugate base form})])/[MOPS(\text{weak acid form})]$$
$$pK_a = -\log_{10}K_a = 7.20$$

If the relative concentrations are known for the weak acid and conjugate base forms of a dissolved weak acid at equilibrium in water, then the solution pH can be determined according to the equation:

$$pH = pK_a + \log_{10}([\text{conjugate base}]/[\text{weak acid}])$$

Therefore,

$$pH = 7.20 + \log_{10}(0.39/0.61) = 7.01$$

What is the pH after 0.008 moles NaOH are added to the MOPS buffer solution? Rather than simply increase the total $[OH^-]$ by 0.008 moles, addition of the strong base shifts the equilibrium of the dissolved MOPS such that [MOPS(weak acid)] *decreases* by 0.008 M and [MOPS(conjugate base)] *increases* by 0.008 M.

Before addition of 0.008 moles NaOH:
$$[MOPS(\text{weak acid})] = 0.61(0.050\ M) = 0.0305\ M$$
$$[MOPS(\text{conjugate base})] = 0.39(0.050\ M) = 0.0195\ M$$

After addition of 0.008 moles NaOH:
$$[MOPS(\text{weak acid})] = 0.0305\ M - 0.008\ M = 0.0225\ M$$
$$[MOPS(\text{conjugate base})] = 0.0195\ M + 0.008\ M = 0.0275\ M$$

The final pH after addition of 0.008 moles NaOH to 50 mM MOPS at pH 7.01 is, therefore:

$$pH = 7.20 + \log_{10}(0.0275/0.0225) = 7.29$$

9. In the acidic pH of a lysosome, ammonia is converted to ammonium ion. Ammonium ion is unable to traverse the membrane because of its positive charge and is trapped within the lysosome. The accumulation of ammonium ion decreases the concentration of protons within lysosomes and therefore elevates lysosomal pH. At neutral pH, ammonia has little, if any, tendency to protonate to ammonium ion and thus has no effect on cytosolic pH.

10. $K_{eq} = [LR]/[L][R]$

Since 90% of L binds R, the concentration of LR at equilibrium is $0.9(1 \times 10^{-3}M) = 9 \times 10^{-4}\ M$. The concentration of free L at equilibrium is the 10% of L that remains unbound, 1×10^{-4} M. The concentration of R at equilibrium is $(5 \times 10^{-2}$ M$)$ − $(9 \times 10^{-4}$ M$) = 4.91 \times 10^{-2}$ M. Therefore, $[LR]/[L][R] = 9 \times 10^{-4}$ M/ $((1 \times 10^{-4}$ M$)$ $(4.91 \times 10^{-2}$ M$)) = 183.3$ M^{-1}.

The equilibrium constant is unaffected by the presence of an enzyme.

$$K_d = 1/K_{eq} = 5.4 \times 10^{-3} \text{ M}.$$

11. $\Delta G = \Delta G^{\circ\prime} + RT\ln [\text{products}]/[\text{reactants}]$

For this reaction, $\Delta G = -1000$ cal/mol $+ [1.987$ cal/(degree $\cdot$ mol) $\times$ (298 degrees) $\times \ln (0.01$ M/$(0.01$ M $\times 0.01$ M$))]$.

$$\Delta G = -1000 \text{ cal/mol} + 2727 \text{ cal/mol} = 1727 \text{ cal/mol}$$

To make this reaction energetically favorable, one could increase the concentration of reactants relative to products such that the term $RT\ln [\text{products}]/[\text{reactants}]$ becomes smaller than 1000 cal/mol. One might also couple this reaction to an energetically favorable reaction.

12. The presence of one or more carbon-carbon double bonds is indicative of an unsaturated or polyunsaturated fatty acid. The term saturated refers to the fact that all carbons, except the carbonyl carbon, have four single bonds. In a cis unsaturated fatty acid, the carbon atoms flanking the double bond are on the same side, thus introducing a kink in the otherwise flexible straight chain. There is no such kink in a trans unsaturated fatty acyl chain.

13. Glutamate is the amino acid that undergoes γ-carboxylation, resulting in the formation of a host of blood clotting factors. Warfarin inhibits γ-carboxylation of glutamate. Thus, blood clotting is severely compromised. Patients prone to forming clots (thrombi) in blood vessels might be prescribed warfarin in order to prevent an embolism, which would result if the clot dislodged and blocked another vessel elsewhere in the body. Patients at risk for heart disease due to blockages in the coronary arteries are also often prescribed this drug.

ANALYZE THE DATA

1. Biomolecules are relatively easy to synthesize from inorganic starting materials, which suggests that living and nonliving matter are not fundamentally different. Living matter is subject to the same laws of physics and chemistry that govern nonliving matter. The fact that biomolecules can be produced through nonliving, chemical processes suggests that life itself could have evolved by similar means. Biochemistry attempts to describe the mechanisms that give rise to living systems from the perspective of the molecules that make up living things. We can often gain considerable insight into the properties of a living thing by studying the structure and chemistry of its molecules.

2. When a weak acid is in aqueous solution of pH at or near its value of pK_a, the weak acid will quickly establish an equilibrium with its conjugate base form and together the two will act to resist additional changes to the solution pH. Solutions in which weak acid/conjugate base pairs function to inhibit pH changes are called "buffers." Buffers are at their most efficient when the concentrations of the weak acid and conjugate base forms are equal, as would be the case when precisely one half the amino acid concentration of sodium hydroxide has been added (0.05 M OH⁻), and at this point the solution pH should equal the pK_a of that weak acid group. At a low pH like 1.8, the buffering species on the amino acid must be a relatively strong type of weak acid like a carboxylic acid. Additional buffer points at pH 6 and 9.3 indicate that there are two additional chemical groups on the dissolved amino acid that can behave as buffers at the appropriate OH⁻ concentrations. The pK_a of 9.3 likely corresponds to the amino group present on every amino acid. The ability of this amino acid to behave like a buffer at three different values of solution pH indicates that the amino acid side chain also has weak acid properties. Its apparent pK_a value of 6.0 identifies this amino acid as histidine due to the fact that its imidazole side chain functions as a buffer at this pH.

3

PROTEIN STRUCTURE AND FUNCTION

REVIEW THE CONCEPTS

1. The primary structure of a protein is the linear arrangement or sequence of amino acids. The secondary structure of a protein is the various spatial arrangements that result from folding localized regions of the polypeptide chain. The tertiary structure of a protein is the overall conformation of the polypeptide chain, its three-dimensional structure. Secondary structures, which include the alpha (α) helix and the beta (β) sheet, are held together by hydrogen bonds. In contrast, the tertiary structure is primarily stabilized by hydrophobic interactions between non-polar side chains of the amino acids and hydrogen bonds between polar side chains. (The quaternary structure describes the number and relative positions of the subunits in a multimeric protein.)

2. Despite the fact that folded proteins adopt conformations that are energetically favorable, the amount of time required for a particular protein to arrive at this conformation on its own can vary significantly. This is especially true if there are other "quasi-stable" conformations available to the polypeptide. Molecular chaperones function to protect an unfolded protein from participating in interactions that will take it off the "pathway" to its native, functional conformation. Chaperonins provide similar support to an unfolded protein. However, chaperonins can also use encapsulation and ATPase activity to give energetic "kicks" to misfolded proteins and get them back on the pathway toward their native folded state.

3. The active site of an enzyme is the region within which the substrate binds and is converted into product. The turnover number (k_{cat}) is the rate constant that can be used to calculate the rate at which a product is formed (V). The Michaelis constant

(K_m) is equal to the substrate concentration at which an enzyme will generate product at precisely one-half its potential maximal velocity. This maximal velocity (V_{max}) is the theoretical limit to the rate at which product can be formed by a particular preparation of enzyme. A rate constant never changes for a particular enzyme. However, the actual measured rate of the chemical reaction will change in response to many variables including the concentration of enzyme, its affinity for substrate, the concentration of substrate, the potential of product to inhibit substrate binding, etc. To calculate a rate (V), one would multiply the turnover number (k_{cat}) and [ES] (the concentration of enzyme-substrate). The rate (V) will equal the maximal rate (V_{max}) when all of the enzyme in a reaction is bound with substrate ([ES] = $[E]_{Total}$).

4. The addition of enzyme does not affect the free energy of either the substrate or the product. Therefore, the difference in free energy (ΔG) for a chemical does not change as a consequence of adding enzyme. The free energy of the enzyme-substrate complex (ES) is lower for E2. Therefore, E2 binds substrate with greater affinity than E1. The transition state ($X^{\ddagger}$) is stabilized equally well by both E1 and E2. Although both E1 and E2 stabilize $X^{\ddagger}$ equally well, E2 binds more tightly to S than E1. This results in a greater activation energy barrier for the reaction catalysed by E2 than by E1 and, therefore, E1 is the better catalyst. In fact, the reaction should proceed at the uncatalyzed rate in the presence of E2 because the height of the barrier between the lowest energy state of the substrate and its transition state is unchanged. By contrast, in the presence of enzyme E1 that barrier is significantly smaller resulting in a more rapid transition between substrate and product.

5. In order for an antibody to catalyze a chemical reaction it should show preferential binding affinity for the transition state of a chemical reaction. As transition states are, by definition, high-energy intermediates of chemical reactions rather than stable molecules, one would first have to synthesize a stable molecule with chemical properties similar to the transition state and use this "transition state analog" as an antigen to promote an adaptive immune response in a test animal.

6. Ubiquitin is a 76–amino acid protein that serves as a molecular tag for proteins destined for degradation. Ubiquitination of a protein involves an enzyme-catalyzed transfer of a single ubiquitin molecule to the lysine side chain of a target protein. This ubiquitination step is repeated many times, resulting in a long chain of ubiquitin molecules. The resulting polyubiquitin chain is recognized by the proteasome, which is a large, cylindrical, multisubunit complex that proteolytically cleaves ubiquitin-tagged proteins into short peptides and free ubiquitin molecules. Proteasome inhibitors would be useful to treat cancers if they blocked the degradation of proteins (e.g., tumor suppressors) required to halt the progression of uncontrolled cell growth. In the case of the proteasome inhibitor Velcade, which is used to treat patients with multiple myeloma, cells undergo apoptosis (programmed cell death), and because a protein serving as a pro-survival factor called NFκB cannot be activated when proteasome activity is blocked (reviewed in A. Fribley and C. Y. Wang, *Cancer Biol. Ther.*, 2006 July 1; 5(7):745–8).

7. Cooperativity, or allostery, refers to any change in the tertiary or quaternary structure of a protein induced by the binding of a ligand that affects the binding

of subsequent ligand molecules. In this way, a multisubunit protein can respond more efficiently to small changes in ligand concentration compared to a protein that does not show cooperativity. The activity of many proteins is regulated by the reversible addition/removal of phosphate groups to specific serine, threonine, and tyrosine residues. Protein kinases catalyze phosphorylation (the addition of phosphate groups), while protein phosphatases catalyze dephosphorylation (the removal of phosphate groups). Phosphorylation/dephosphorylation changes the charge on a protein, which typically leads to a conformational change and a resulting increase or decrease in activity. Some proteins are synthesized as inactive propeptides, which must be enzymatically cleaved to release an active protein.

8. Proteins can be separated by mass by centrifuging them through a solution of increasing density, called a density gradient. In this separation technique, known as rate-zonal centrifugation, proteins of larger mass generally migrate faster than proteins of smaller mass. However, this is not always true because the shape of the protein also influences the migration rate. Gel electrophoresis can also separate proteins based on their mass. In this technique, proteins are separated through a polyacrylamide gel matrix in response to an electric field. Because the migration of proteins through a polyacrylamide gel is also influenced by shape of proteins, the ionic detergent sodium dodecyl sulfate is added to denature proteins and force proteins into similar conformations. During rate-zonal centrifugation, a protein of larger mass (transferrin) will sediment faster during centrifugation, whereas a protein of smaller mass (lysozyme) will migrate faster during electrophoresis.

9. Gel filtration, ion exchange, and affinity chromatography typically involve the use of a bead consisting of polyacrylamide, dextran or agarose packed into a column. In gel filtration chromatography, the protein solution flows around the spherical beads and interacts with depressions that cover the surface of the beads. Small proteins can penetrate these depressions more readily than larger proteins and thus spend more time in the column and elute later from the column; larger proteins do not interact with these depressions and elute first from the column. In ion-exchange chromatography, proteins are separated on the basis of their charge. The beads in the column are covered with amino or carboxyl groups that carry a positive or negative charge, respectively. Positively charged proteins will bind to negatively charged beads, and negatively charged proteins will bind to positively charged beads. In affinity chromatography, ligand molecules that bind to the protein of interest are covalently attached to beads in a column. The protein solution is passed over the beads and only those proteins that bind to the ligand attached to the beads will be retained, while other proteins are washed out. The bound protein can later be eluted from the column using an excess of ligand or by changing the salt concentration or pH.

10. Proteins can be made radioactive by the incorporation of radioactively labeled amino acids during protein synthesis. Methionine or cysteine labeled with sulfur-35 are two commonly used radioactive amino acids, although many others have also been used. The radioactively labeled proteins can be detected by autoradiography. In one example of this technique, cells are labeled with a radioactive compound and then overlaid with a photographic emulsion sensitive to

radiation. The presence of radioactive proteins will be revealed as deposits of silver grains after the emulsion is developed. A Western blot is a method for detecting proteins that combines the resolving power of gel electrophoresis, the specificity of antibodies, and the sensitivity of enzyme assays. In this method, proteins are first separated by size using gel electrophoresis. The proteins are then transferred onto a nylon filter. A specific protein is then detected by use of an antibody specific for the protein of interest (primary antibody) and an enzyme-antibody conjugate (secondary antibody) that recognizes the primary antibody. The presence of this protein-primary antibody-enzyme-conjugated secondary antibody complex is detected using an assay specific for the conjugated enzyme.

11. X-ray crystallography can be used to determine the three-dimensional structure of proteins. In this technique, x-rays are passed through a protein crystal. The diffraction pattern generated when atoms in the protein scatter the x-rays is a characteristic pattern that can be interpreted into defined structures. Cryoelectron microscopy involves the rapid freezing of a protein sample and examination with a cryoelectron microscope. A low dose of electrons is used to generate a scatter pattern that can be used to reconstruct the protein's structure. In nuclear magnetic resonance (NMR) spectroscopy, a protein solution is placed in a magnetic field and the effects of different radio frequencies on the spin of different atoms are measured. From the magnitude of the effect of one atom on an adjacent atom, the distances between residues can be calculated to generate a three-dimensional structure.

    X-ray crystallography can provide extremely high-resolution structural information on molecules and molecular complexes of any size. The principal disadvantage of x-ray crystallography is the challenge of producing samples in the form of single crystals suitable for diffraction experiments. NMR spectroscopy gives high-resolution information on protein structures in solution. It also is ideal for monitoring protein dynamics. However, NMR spectroscopy is limited in its ability to conclusively determine the structures of very large proteins and symmetrical macromolecular assemblies. The principal advantage of electron microscopy is the relative ease of sample preparation. However, structural resolution is generally not so high as with the other methods, especially for asymmetric assemblies. NMR is better for small proteins. Electron microscopy and x-ray crystallography, so long as a suitable crystal can be obtained, are ideal for large proteins and macromolecular assemblies.

12. The four features of a mass spectrometer are 1) the ion source, 2) the mass analyzer, 3) the detector, and 4) a computerized data system. Basically, the investigator would collect protein samples from the cancerous cells and from the normal healthy cells, the latter serving as a control. Samples would be prepared for 2D PAGE and after electrophoresis the gels would be dyed and the profiles compared. If a protein "spot" were present in the sample from the cancer cell and not the control, it would be isolated out of the gel, protease-digested using trypsin to generate peptides that are mixed with a matrix, and applied to a metal target. A laser is used to ionize the peptides, which are vaporized into singly charged ions. In the case of a time of flight (TOF) mass analyzer, the time it takes the ions to pass through the analyzer before reaching

the detector is inversely proportional to its mass and directly proportional to the charge they carry, generating a spectrum in which each molecule has a distinct signal, allowing the investigator to calculate each ion's mass. The fourth essential component is a computerized data system that acquires and stores the data, which are then compared to information in databases. The mass and charge signature, or fingerprint, of the unknown is compared to that of peptides in a database and the best match protein is identified.

ANALYZE THE DATA

2. a. Proteins 3, 5, 6, and 7 do not change in response to the drug. Protein 1 declines in response to the drug. Proteins 2 and 4 are induced in response to the drug.

b. Proteins 2, 3, and 6 are phosphoproteins; the others are not. Protein 2 is a phosphoprotein normally present in cells whose level increases in response to the drug. Protein 3 is not a phosphoprotein in control cells and is phosphorylated in response to the drug. Protein 6 is a phosphoprotein whose level does not change in response to the drug.

c. Proteins 1 and 6 are strictly nuclear proteins. Proteins 2, 3, and 4 are strictly cytoplasmic proteins. Protein 5 is present in the nucleus and cytoplasm. Protein 7 is a cytoplasmic protein that migrates to the nucleus in response to drug treatment.

d. Protein 1 is a nuclear protein whose level declines in response to drug treatment. Protein 2 is a cytoplasmic phosphoprotein whose level increases in response to drug treatment. Protein 3 is a cytoplasmic phosphoprotein whose level does not change in response to drug treatment. Protein 4 is a cytoplasmic protein whose level increases in response to drug treatment. Protein 5 is a nuclear and cytoplasmic protein whose level does not change in response to drug treatment. Protein 6 is a nuclear phosphoprotein whose level does not change in response to drug treatment. Protein 7 is a cytoplasmic protein which migrates from the cytoplasm to the nucleus, in response to drug treatment.

4

BASIC MOLECULAR GENETIC MECHANISMS

REVIEW THE CONCEPTS

1. Watson-Crick base pairs are interactions between a larger purine and a smaller pyrimidine base in DNA. These interactions result in primarily G-C and A-T base pairing in DNA and A-U base pairs in double-stranded regions of RNA. They are important because they allow one strand to function as the template for synthesis of a complementary, antiparallel strand of DNA or RNA.

2. At 90°C, the double-stranded DNA template will denature and the strands will separate. As the temperature slowly drops below the Tm of the plasmid DNA, the single-stranded oligonucleotide primer present at higher concentration than the plasmid DNA strands hybridizes to its complementary sequence on the plasmid template. The resulting molecules contain a short double-stranded stretch the length of the primer with a free 3′ OH that can be used by DNA polymerase enzyme in sequencing reactions.

3. RNA is less stable chemically than DNA because of the presence of a hydroxyl group on C-2 in the ribose moieties in the backbone. Additionally, cytosine (found in both RNA and DNA) may be deaminated to give uracil. If this occurs in DNA, which does not normally contain uracil, the incorrect base is recognized and repaired by cellular enzymes. In contrast, if this deamination occurs in RNA, which normally contains uracil, the base substitution is not corrected. Thus, the presence of deoxyribose and thymine make DNA more stable and less subject to spontaneous changes in nucleotide sequence than RNA. These properties might explain the use of DNA as a long-term information-storage molecule.

4. In prokaryotes, many protein-coding genes are clustered in operons where transcription proceeds from a single promoter that gives rise to one mRNA encoding multiple proteins with related functions. In contrast, eukaryotes do not have operons but do transcribe intron sequences that must be spliced out of mature mRNAs. Eukaryotic mRNAs also differ from their prokaryote counterparts in that they contain a 5' cap and 3' poly(A) tail. Also, ribosomes have immediate access to nascent mRNAs in bacteria so that translation begins as the mRNA is being synthesized. In contrast, in eukaryotes, mRNA synthesis occurs in the nucleus, whereas translation by ribosomes occurs in the cytoplasm. Consequently, only fully synthesized and processed mRNAs are translated in eukaryotes.

5. A simple explanation is that the larger, membrane-spanning, domain-containing protein and the small, secreted protein are encoded by the same gene that is differentially spliced. Specifically, the final exon of the gene could contain the information for the membrane-spanning domain, and in the smaller, secreted protein, this exon could be omitted during splicing.

6. An operon is an arrangement of genes in a functional group that are devoted to a single metabolic purpose. In the case of tryptophan synthesis, the DNA for five genes is arranged in a contiguous array that gets transcribed from a single promoter into a continuous strand of mRNA encoding five proteins. In this manner, the cell simply has to induce one promoter, which transcribes all the necessary genes encoding the proteins (enzymes) to make the amino acid tryptophan. Splicing out intronic sequences or transcribing multiple mRNAs from genes on different chromosomes, as seen in eukaryotic systems, is unnecessary; thus, operons are a logical way to economize on the amount of DNA needed by genes to encode a number of proteins. In addition, this arrangement allows all the genes in an operon to be coordinately regulated by controlling transcription initiation from a single promoter.

7. Since poly(A)-binding protein is involved in increasing the efficiency of translation, a mutation in poly(A)-binding protein would cause less efficient translation. Polyribosomes in a cell with such a mutation would not contain circular structures of mRNAs during translation because lack of the poly(A)-binding protein would eliminate the 3' binding site for eIF4G.

8. DNA synthesis is discontinuous because the double helix consists of two antiparallel strands and DNA polymerase can synthesize DNA only in the 5' to 3' direction. Thus, one strand is synthesized continuously at the growing fork, but the other strand is synthesized utilizing Okazaki fragments that are joined by DNA ligase.

9. Base excision repair is responsible for repairing guanine-thymine mismatches caused by the chemical conversion of cytosine to uracil or by deamination of 5-methyl cytosine to thymine. Mismatch excision repair eliminates base pair mismatches and small insertions or deletions of nucleotides generated accidentally during DNA replication. Nucleotide excision-repair fixes DNA strands that contain chemically modified bases, which ensures that thymine-thymine dimers are repaired in the case of UV light damage.

10. UV irradiation causes thymine-thymine dimers. These are usually repaired by the nucleotide excision-repair system, which utilizes XP complexes and the transcriptional helicase TFIIH to unwind and excise the damaged DNA. The gap is then filled in by DNA polymerase. Ionizing radiation causes double-stranded breaks in DNA. Double-stranded breaks are repaired either by homologous recombination or nonhomologous DNA end-joining. Homologous recombination requires the BRCA1, BRCA2, and Rad51 proteins to use the sister chromatid as template for error-free repair. Nonhomologous DNA end-joining is error-prone because nonhomologous ends are joined together. Since formation of a malignant tumor requires multiple mutations, cells that have lost DNA-repair function are more likely to sustain cancer-promoting mutations. Examples are xeroderma pigmentosum due to mutations in *XP* genes that prevent repair of thymine dimers and a genetic predisposition to breast cancer in individuals with germ-line mutations in the *BRCA1* or *BRCA2* genes.

11. Homologous recombination is the process that can repair DNA damage and also generate genetic diversity during meiosis. In both cases, repair is to double-strand breaks, RecA/Rad51-like proteins play key roles in the recombination process, and Holliday structures form, followed by cleavage and ligation to form two recombinant chromosomes. During DNA repair by homologous recombination, the damaged sequence is copied from an undamaged copy of the homologous DNA sequence on the homologous chromosome or sister chromatid. During meiosis, however, genetic diversity is generated by homologous recombination where large regions of chromosomes are exchanged between the maternal and paternal pair of homologous chromosomes. Also, in meiosis an exchange called crossing over is required for the proper segregation of the chromosomes during the first meiotic cell division.

12. The gene encoding the reverse transcriptase enzyme is unique in retroviruses and closely related retrotrasposons. These viruses contain RNA as their genetic material; a DNA copy of the viral RNA is made during infection and reverse transcriptase catalyzes this reaction. The human T-cell lymphotrophic virus, which causes T-cell leukemia, and human immunodeficiency virus, which causes AIDS, can infect only specific cell types because these cells possess receptors that interact specifically with viral envelope proteins of the progeny virus.

13. a. bottom strand

    b. 5'ACGGACUGUACCGCUGAAGUCAUGGACGCUCGA 3'

14.

Prokaryotes	Eukaryotes
Very little non-coding DNA	Non-coding DNA (introns) interspersed between coding regions (exons)
Genes that carry out similar/complementary functions are in tandem on the chromosome (operon).	Genes with similar/complementary functions are interspersed throughout the chromosome; some are on different chromosomes.

(cont.)

Direct production of mRNA (no processing)	mRNA produced after processing of primary RNA to remove introns, 5'-methyl capping and 3'-polyadenylation
Ribosomes have immediate access to mRNA to initiate protein synthesis (no nucleus).	mRNA has to be translocated from the nucleus to cytoplasm before protein synthesis can begin.
One mRNA $\longrightarrow$ many polypeptides	One mRNA $\longrightarrow$ one polypeptide

15. a. Double-stranded DNA won't be unwound long enough to allow for the replication of the DNA.

   b. Translation will not occur because the initiation complex will be unable to bind the AUG start site.

   c. Translation will not occur because tRNA$_i$ ^{met} is the only tRNA able to initiate translation.

16. RNA sequence: 5′ UUC UAC AUG AAG CAU CAG AGC CAG UGA 3′

   Protein sequence:

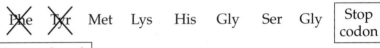

   Phe Tyr Met Lys His Gly Ser Gly | Stop codon |

   Not produced because they are before the start codon

17. a. Synthesis is from 5′ $\longrightarrow$ 3′; as the DNA strand separates, one strand serves as an anti-parallel template whereby more and more of the template for the 3′ end of the newly synthesized strand is revealed as the replication fork advances making the synthesis from 5′ $\longrightarrow$ 3′ a continuous process. At the same time, the template for the 5′ end of the other strand is continuously revealed as the fork advances, therefore small 5′ $\longrightarrow$ 3′ fragments need to be synthesized on this strand.

   b. If the replication fork were moving in the opposite direction (left to right) to the example shown above.

18. The newly synthesized strand is the one with the error—if the original strand were targeted, the mutation would be allowed to persist and would be transmitted to all subsequent cells.

19. a. nonsense (TCA $\longrightarrow$ TGA)

   b. missense AUG $\longrightarrow$ AUA (methionine $\longrightarrow$ isoleucine)

20. a.

Lytic	Non-lytic
Infected cell ultimately dies.	Infected cell does not die.
Viral genome does not integrate into host genome; host cell DNA destroyed.	Viral genome integrates into host genome.

   b. ii. Viral mRNAs are transcribed by the host-cell translation machinery.

ANALYZE THE DATA

   a. The data suggest that context matters and that a change in the sequence surrounding the first AUG affects the efficiency of initiation from this start site. When comparing lanes 1 and 2, in which the mRNAs differ only at position (+4), one observes that a G, rather than a U, at this position reduces leaky scanning. More preCAT and less CAT is synthesized with the message used in lane 2 than with that used in lane 1. Although (−3)ACCAUGG(+4) is hypothesized to provide an optimal context in which the first AUG is presented, the data suggest that this sequence can be modified to ACCAUGA without significantly compromising efficiency of initiation from the first AUG. The only difference between the messages in lanes 3 and 4 is a change from G to A at position (+4), and in each case only preCAT is synthesized. With respect to the importance of ACC at positions (−3) to (−1), a comparison of lanes 4 and 5 reveal that a shift of ACC from position (−3) to (−1) (lane 4) to position (−4) to (−2) (lane 5) results in a loss of fidelity of initiation from the first start site. Replacement of (−3)ACC(−1) with (−3)UUU(−1) (compare lane 2 to lane 3) results in a loss of efficiency of translation from the first start site, so these data suggest that the ACC sequence and its position relative to the AUG matter.

   b. In order to further test the importance of the nucleotide at the +4 position, it would be useful to undertake an analysis of CAT mRNA mutants with only single substitutions compared to the wild-type sequence shown in lane 3. The data shown provide evidence that G can be substituted with A (lane 4) at position (+4) and suggest that it cannot be substituted with U (lane 1). However, there are additional sequence changes in the mRNA in lane 1 other than the change at position (+4). It would be informative to change the mRNA used in lane 3 so that either U or C is substituted for G at position (+4). If each of these substitutions results in synthesis of some CAT, then one could deduce that an optimal context for the first start site can tolerate purines but not pyrimidines at the (+4) position. The data shown in the figure also do not examine the importance of the (−3)ACC(−1) sequence other than to change it completely to UUU. It would also be useful to change this sequence one nucleotide at a time to determine the relative importance of each of these nucleotides in helping the ribosomes pause at the first AUG site and begin translation. In each case, synthesis of CAT would be evidence that the particular substitution to the sequence results in a loss of fidelity of initiation at the first AUG. If the A at position (−3) is the most important of the ACC sequence for generating efficiency of translation for the first AUG, then one would expect that changes to this nucleotide would result in synthesis of more CAT than would changes to the other two nucleotides.

c.   The mutation in this family results in the introduction of a new AUG sequence in the hepcidin mRNA upstream from the original start site. Because the new AUG is not in frame with respect to the original start AUG, no hepcidin will be made if initiation begins exclusively at this new, upstream AUG. The fact that hepcidin is not synthesized in individuals who have inherited this mutation suggests that initiation of protein synthesis occurs at the new AUG with high efficiency and that the ribosomes do not scan through this site to begin synthesis downstream at the original start site. Thus, these findings support the hypothesis that initiation of protein synthesis in eukaryotes normally begins at the first 5′ AUG site. An examination of the context in which this new AUG start site is located reveals that it has the important G located at (+4), as does the original AUG start site. Whereas the original AUG has an A at the (−3) position (and, in fact, also has the consensus C at position (−2)), the new, upstream AUG does not have the consensus sequence in this position. However, given that no hepcidin is made, the new start site may be in a context that facilitates efficient recognition by the ribosomes and thereby does not result in any (or in any detectable) leaky scanning.

5

MOLECULAR GENETIC TECHNIQUES

REVIEW THE CONCEPTS

1. A recessive mutation must be present in both alleles of a diploid organism in order for the mutant phenotype to be observed; that is, the individual must be homozygous for the mutation to be expressed. Recessive alleles usually result from a mutation that inactivates the affected gene, leading to loss of function. If the inactivated gene is an essential gene, loss of function would be lethal in homozygotes with both mutated alleles. Lethal recessive mutations can be maintained in heterozygotes. In contrast, a dominant mutation produces a mutant phenotype even in the presence of one mutant and one wild-type allele. Dominant alleles often result from a mutation that causes some kind of gain of function. Dominant alleles that affect the function of essential genes can be lethal even in heterozygotes. Thus, researchers may use conditional mutations such as temperature sensitive mutations to study the effects of dominant lethal alleles.

2. A temperature sensitive mutation is one where the gene is non-functional at a given temperature, usually a lower or higher than normal growth temperature. These are useful for the study of essential genes, as it is not possible to create a viable cell that lacks an essential gene.

3. Complementation analysis can be used to determine whether two recessive mutations are present in the same or different genes. If a heterozygous organism containing both mutations shows the mutant phenotype, then the two mutations are in the same gene because neither allele provides a functional copy of the gene. In contrast, if a heterozygous organism shows a wild-type phenotype, then the two mutations are in

different genes because a wild-type allele of each gene is present. Dominant mutations cannot be tested by complementation analysis because they will display a mutant phenotype even in the presence of a wild-type allele of the gene.

4. She will grow them on media without lysine. These yeast cannot normally grow without lysine, although when they acquire a library plasmid they will be capable of synthesizing it. She will know her mutant has been complemented when she sees a colony that is now white (the wild-type color).

5. Bacteria that synthesize restriction enzymes also synthesize a DNA modifying enzyme to protect its own DNA. The modifying enzyme is a methylase, which methylates the host DNA. Methylated DNA is no longer a substrate for the encoded restriction enzyme. Restriction enzyme sites commonly consist of 4–8 base pair, palindromic sequences. After being cut with a restriction enzyme, the ends of the cut DNA molecule can exist as single-stranded tails (sticky ends) with either 5' or 3' overhangs, or as blunt (flush) ends. DNA ligase is an enzyme that catalyzes the reformation of the phosphodiester bond between nucleotides in the presence of ATP.

6. A plasmid is a circular, extrachromosomal DNA molecule that contains an origin of replication, a marker gene that permits selection, and a region into which foreign DNA can be inserted (cloning site). A plasmid is useful for cloning DNA fragments up to approximately 20 kb. Specialized plasmid vectors such as BACs (bacterial artificial chromosomes) have been developed that can accommodate DNA fragments as large as several million nucleotides.

7. DNA libraries are collections of randomly cloned DNA fragments. A cDNA (complementary DNA) library is a collection of DNA molecules that are copied from messenger RNA molecules using the enzyme reverse transcriptase. A cDNA library does not contain every gene, only those that are expressed as mRNA at the time of RNA isolation. In contrast, a genomic DNA library consists of random fragments of the total genome. This would include not only genes but also areas of the genome that do not encode for genes. You could use either of the genomic libraries or the cDNA library from neurons. You could not use the cDNA skin library because cDNA libraries are based on the cell's mRNA. If the skin cell does not express Gene X, that cDNA will not be present in the library.

8. The PCR reaction is performed as multiple cycles of a three-step process. The first step involves heat denaturation of a target DNA molecule. The second step involves cooling the DNA solution to allow annealing of short single-stranded oligonucleotide primers that are complementary to the target DNA molecule. In the final step, the hybridized oligonucleotides serve as primers for DNA synthesis. The resultant double-stranded DNA molecules are then subjected to further rounds of denaturation, annealing, and DNA synthesis (extension). A thermostable DNA polymerase was essential for automation of the PCR process. A nonthermostable DNA polymerase would be inactivated by heat denaturation during each cycle of the PCR process and would necessitate the addition of new enzyme prior to each DNA synthesis step.

9. Southern blotting is a technique in which DNA fragments are separated by size in a gel and then transferred to a solid support such as a nitrocellulose or nylon membrane. The DNA is fixed to the nylon membrane and hybridized to a labeled DNA or RNA probe. The hybridized probe is then detected by some technique such as autoradiography. Northern blotting is similar to Southern blotting, except that RNA instead of DNA is denatured and then separated on the gel. Southern blotting can be used to identify a DNA fragment that contains a DNA sequence of interest. Northern blotting can be used to determine the steady-state levels of a specific RNA.

10. In order to express a foreign gene, a recombinant plasmid would require a promoter for efficient transcription of the foreign gene. A promoter that is inducible would provide even higher expression levels of the foreign gene product. To facilitate purification of the foreign protein, a molecular tag can be added to the recombinant protein. An example of this type of molecular tag is a short sequence of histidine residues (a polyhistidine sequence). The resultant His-tagged protein will bind specifically to a bead that has bound nickel atoms. Other proteins can be washed out and the His-tagged protein can be released from the nickel atoms by lowering the pH of the solution. Bacterial cells are limited in their capacity to synthesize complex proteins because of their inability to perform many post-translational modifications, such as glycosylation, that mammalian cells can perform. These posttranslational modifications are essential for the biological activity of the recombinant protein.

11. Northern blotting and RT-PCR are useful for analyzing fewer genes because they rely on the creation of specific radioactive probes or PCR primers. It is less feasible to make large quantities of these for whole genome analysis. Studying entire genomes (or chromosomes in this example) is most easily accomplished with a microarray, which uses a chip that can hybridize to hundreds or thousands of genes based on complementary base pairing.

12. The expression of mRNA in individual cells can be determined by in situ hybridization in whole cells or tissue sections. Fixed cells are exposed to labeled DNA probes that are complementary to the mRNA of interest. After washing to remove excess probe, the cells can be examined microscopically to detect the locations of labeled mRNA. This process can also be used to identify mRNA locations in embryos.

13. Single-nucleotide polymorphisms (SNPs) are changes in a single nucleotide between two individuals. Simple-sequence repeats (SSRs), also known as microsatellites, consist of a variable number of repeating one-, two-, or three-base sequences. The number of these repeat units at a specific genetic locus varies between individuals. Both types of polymorphisms can be used as molecular markers for mapping studies. The recombination frequency between two polymorphisms can be determined and can serve as the basis for development of a genetic map. In general, the farther two markers are separated on a chromosome, the greater the recombination frequency between those two markers, and vice versa.

14. Linkage disequilibrium mapping can sometimes be used in cases where a genetic disease commonly found in a particular population results from a single mutation that occurred many generations in the past. In such cases, most of the individuals with the disease would have inherited the disease from the same ancestral chromosome. The closer genetic markers are to each other, the less likely they will be recombined by crossing over during meiosis. Thus, individuals inherit sections of DNA from their parents, not just individual genes. DNA polymorphisms on part of a chromosome that are inherited together are called haplotypes. If geneticists can identify a haplotype common to all the affected individuals in a particular population, DNA markers associated with the disease haplotype might help localize the disease-associated gene to a relatively small chromosomal region.

15. Once a gene is roughly located along a chromosome by genetic linkage studies, further analysis is required to identify the "disease" gene. One strategy for identifying a disease gene involves gene expression analysis. Comparison of gene expression in tissues from normal and affected individuals by Northern blot analysis may indicate a gene that is involved in the disease process. Northern blot analysis allows a comparison of both the level of expression and the size of the transcripts between normal and disease tissues. Sometimes expression levels and/or size of the transcripts do not alter between the normal and the disease states. In this case, DNA sequencing of a potential disease gene from tissues of a normal and a disease state could reveal a single nucleotide change that results in the disease phenotype.

16. To generate a knockout mouse, mouse embryonic stem cells are first transfected with a disrupted allele of the target gene. Through a process known as homologous recombination the disrupted allele replaces the functional homologous gene in the chromosome, resulting in a nonfunctional chromosomal gene. The ES cells, which now contain a mutant gene, are injected into a blastocyst. The blastocyst is transferred into a recipient mouse. Pups that are born will be chimeras. The loxP-Cre system can be used to conditionally knock out a gene. Using the above technology, loxP sites can be engineered to flank the gene of interest. Expression of the recombinase, Cre, in a specific tissue will result in loss of the flanked gene in that tissue. Knockout mice serve as models for human diseases. For example, if a human disease is known to result from a mutation in gene X, a knockout mouse can be generated that lacks gene X.

17. A dominant negative mutation is a mutation that produces a mutant phenotype even in cells carrying a wild-type copy of the gene. This type of mutation produces a loss of function phenotype. RNA interference (RNAi) is a method of inactivating gene expression by selectively destroying RNA. In this method, a short double-stranded RNA molecule is introduced into cells. This double-stranded RNA base pairs with its target mRNA, promoting degradation of the mRNA by specific nucleases.

ANALYZE THE DATA

a. X and Y cells do not grow at the elevated temperature, indicating mutations X and Y are temperature sensitive mutations that are present in essential genes. Furthermore, the X and Y mutations complement each other, as shown by the growth of the X-Y diploid at the elevated temperature. This indicates that mutations X and Y are in different genes. It also tells us that each mutation is recessive, as a single copy of the wild-type gene is sufficient to overcome the growth defect.

b. The plate on the left lacks uracil; clones growing on this plate must be able to synthesize their own uracil (they must have a wild-type copy of the *URA3* gene). The yeast themselves are defective in uracil synthesis, so each clone on the left plate must be transformed with a plasmid containing the *URA3* gene. The single clone on the plate at the right grew in the absence of uracil at the restrictive temperature. Therefore, it must harbor a plasmid that contains the wild-type copy of the *X* gene, allowing mutant X cells to grow at 32°C. If the plasmid were to be re-isolated from this clone and the yeast cDNA insert analyzed and sequenced, the *X* gene will have been identified. Extract the plasmid from the complemented yeast and sequence it. The sequencing primer will be complementary to the vector backbone.

c. The Southern data reveal that the DNA restriction fragment that contains gene *X* is larger in mutant X than in the parental strain or in mutant Y, which both contain wild-type copies of gene *X*. This larger restriction fragment could arise in a number of ways. An alteration in the fragment size could be due to a mutation that affects the sequence of a restriction site. Alternatively, an insertion of DNA sequence into gene *X* could lead to a larger restriction fragment. However, the PCR data indicate that gene *X* is the same size in mutant X as it is in the parental strain. Accordingly, these data suggest that the mutation in X is a result of a base change(s) in gene *X*, which, fortuitously, has resulted in the loss of a restriction site.

d. At 32°C, mutant X cells containing the wild-type X-GFP fusion construct do not grow—in other words, presence of X-GFP protein is unable to rescue growth at the restrictive temperature. Even though it is synthesized, X-GFP protein cannot compensate for the mutation in X. The GFP-X construct, on the other hand, is able to rescue growth of X cells at the restrictive temperature. The fluorescent microscopy data show that the GFP-X protein is localized to the nucleus, while the X-GFP protein is localized to the cytoplasm. The presence of GFP at the C terminus of X seems to interfere with protein X's ability to be localized to the nucleus. Furthermore, it appears that X must be localized to the nucleus in order to function.

e. The gene product of X is needed earlier in the process of bud formation (earlier in the cell cycle) than the gene product of Y.

6

GENES, GENOMICS, AND CHROMOSOMES

REVIEW THE CONCEPTS

1. A gene is commonly defined as the entire nucleic acid sequence that is necessary for the synthesis of a functional gene product (RNA or polypeptide). This definition includes introns and the regulatory regions (e.g., promoters and enhancers of the gene).

    (i) a, b, c

    (ii) b, c

    (iii) a

2. Single or solitary genes are present once in the haploid genome. In multicellular organisms, roughly 25–50% of the protein-coding genes are solitary genes. Gene families are sets of duplicated genes that encode proteins with similar but nonidentical amino acid sequences. An example of a gene family is the β-like globin genes. Pseudogenes are copies of genes that are nonfunctional even though they seem to have the same exon-intron structure as a functional gene. Pseudogenes probably arise from a gene duplication followed by the accumulation of mutations that render the gene nonfunctional. Tandemly repeated genes are present in a head-to-tail array of exact or almost exact copies of genes. Examples of tandemly repeated genes include ribosomal RNAs and RNAs involved in RNA splicing.

3. Satellite, or simple sequence, DNA can be categorized as microsatellite or minisatellite DNA depending upon the size of the repeated DNA sequence. Microsatellites contain repeats that contain 1–13 base pairs. Minisatellites consist of repeating units

of 14–100 base pairs, present in relatively short regions of 1 to 5 kb made up of 20–50 repeat units. The number of copies of the tandemly repeated DNA sequences varies widely between individuals. DNA fingerprinting is a technique that examines the number of repetitive units at a specific genetic locus for several separate loci. This technique can distinguish and identify individuals on the basis of differences in the number of repeats in their micro- and minisatellite simple-sequence DNA.

4. A bacterial insertion sequence, or IS element, is a member of the class of mobile DNA elements. The IS element usually contains inverted repeats at the end of the insertion sequence. Between the inverted repeats is a region that encodes the enzyme transposase. Transposition of the IS element is a three-step process. First, transposase excises the IS element in the donor DNA; second, it makes staggered cuts in a short sequence in the target DNA; and, third, it ligates the 3′ termini of the IS element to the 5′ ends of the cut donor DNA. The final step involves a host-cell DNA polymerase, which fills in the single-stranded gaps, generating 5–11 base pair short direct repeats that flank the IS element before DNA ligase joins the free ends.

5. Retrotransposons transpose through an RNA intermediate. One class of retro-transposons, the LTR retrotransposons, contain long terminal repeats (LTRs) at their ends and a central protein coding region that encodes the enzymes reverse transcriptase and integrase. The retrotransposon is first transcribed into RNA by host RNA polymerase. This RNA intermediate is then converted into DNA by the action of reverse transcriptase primed with a cellular tRNA in the cytoplasm. The double-stranded DNA copy generated is then imported into the nucleus and inserted into chromosomal DNA by the action of an integrase that is similar to the transposases of DNA transposons. Retrotransposons that lack LTRs transpose by a different mechanism. Non-LTR retrotransposons, of which LINES are an example, consist of direct repeats that flank a region that encodes two proteins: ORF1, an RNA-binding protein, and ORF2, which is similar to reverse transcrip-tase. The LINE element is first transcribed by host RNA polymerase and exported to the cytoplasm, resulting in the translation of ORF1 and ORF2. These proteins bind the LINE RNA and import it into the nucleus, where ORF2 makes staggered nicks in A/T-rich target DNA. The resulting T-rich strand of chromo-somal DNA then hybridizes to the poly (A) tail at the 3′-end of the LINE RNA and primes reverse transcription of the RNA by ORF2 protein. The RNA strand of the resulting RNA/DNA hybrid is replaced by DNA and both resulting DNA strands are ligated to the chromosome ends generated by the original ORF2 cut through the action of host-cell enzymes that normally replace the RNA primers of Okazaki fragments and ligate them together during cellular DNA synthesis.

6. Insertion of transposons can generate spontaneous mutations that may influence evolution. In addition, unequal crossing over between homologous mobile ele-ments at different chromosomal locations leads to exon duplications, gene dupli-cations, and chromosomal rearrangements that can generate new combinations of exons. Subsequent divergence of duplicated genes leads to members of gene families with distinct functions. The inclusion of flanking DNA during transposi-tion also results in the movement of genomic DNA to another region of the

genome. This can result in new combinations of exons, an evolutionary process known as exon shuffling, as well as new combinations of transcriptional control regions.

7. Similarities between bacteria, mitochondria, and chloroplasts reflect the proposed endosymbiotic origin of mitochondria and chloroplasts. Mitochondrial and bacterial ribosomes resemble each other and differ from eukaryotic cytosolic ribosomes in their RNA and protein compositions, their size, and their sensitivity to antibiotics. Bacterial and mitochondrial ribosomes are sensitive to chloramphenicol but resistant to cycloheximide. Eukaryotic cytosolic ribosomes are sensitive to cycloheximide and resistant to chloramphenicol. Also, comparing the mitochondrial DNA of multiple classes of eukaryotes, both unicellular and multicellular, all can be seen to derive from a common ancestor with a genome similar to contemporary symbiotic bacteria that invade host eukaryotic cells. Mitochondrial DNA in different contemporary eukaryotes can be derived from this common ancestor by deletion of different sets of genes in the mitochondria of different eukaryotes and the transfer of genes essential for mitochondrial function to the nucleus.

8. Paralogous genes are genes that have diverged as a result of a gene duplication (i.e., two genes in an organism that have different functions but very similar nucleotide sequences). Orthologous genes are genes that arose because of speciation (i.e., genes found in different species that have very similar nucleotide sequences and functions). Because of alternative splicing, a gene can give rise to numerous protein products. Thus, a small increase in gene number could result in a very large increase in protein number. Thus, the number of proteins and protein-protein interactions could be much greater in the organism with the larger genome. Also, complexity among multicellular organisms arises largely from organizing a larger number of cells into more complex groups of interacting cells. This requires evolution of control regions that regulate transcription and cell division as well as the evolution of new proteins.

9. A nucleosome consists of a protein core of histones with DNA wound around its surface. The protein core consists of an octamer, containing two copies of histones H2A, H2B, H3, and H4. Approximately 150 base pairs of DNA are wrapped less than two complete turns around the octameric histone core. The histone H1 binds to the linker region, which varies in length from 10 to 90 base pairs and is located between nucleosomes. The nucleosomes are folded into a two-start helix (see Figure 6-30) to form a 30-nm fiber.

10. Chromatin modifications affect whether the DNA is tightly or loosely compact. As transcription factors and RNA polymerase must be able to access the DNA, transcription will only occur when the DNA is loosely compacted. When a gene is actively transcribed, the chromatin would likely be acetylated. When a gene is not actively transcribed, the chromatin would likely be methylated or deacetylated.

11. A eukaryotic chromosome consists of a long, linear DNA molecule. The DNA is wrapped around octameric histone cores to form chromatin, which is then

condensed into a 30-nm fiber. Long loops of the 30-nm fiber are thought to be tethered at the base by SMC proteins that encircle chromatin fibers (see Figure 6-36). Scaffold associated regions (SARs) or matrix attachment regions (MARs) are regions in the DNA thought to be associated with the bases of these loops. Genes are primarily located within chromatin loops (i.e., between SARs or MARs). DNA needs to be decondensed for transcription to occur. Decondensing chromatin in the SARs region would cause the chromatin scaffold to disassemble.

12. FISH, or fluorescent in situ hybridization, is one of many related techniques used to detect DNA (or RNA) sequences in cells or tissues. FISH involves the addition of a fluorescent probe that has the sequence complementary to a specific region on the chromosome. Upon addition to the chromosomes, the probe will bind to the DNA and be visible under a fluorescent microscope. In the case of DNA, a fluorescent probe to a specific sequence is made and then hybridized to its complementary sequence directly on the chromosome(s). The signal is detected using fluorescence microscopy. Multicolor FISH is used to detect chromosomal translocations. For example, patients with chronic myelogenous leukemia possess leukemic cells with the Philadelphia chromosome, a shortened chromosome 22 containing genetic material translocated from chromosome 9. These patients also have an abnormally long chromosome 9, resulting from a considerable amount of DNA that translocated from the long arm of chromosome 22. By labeling leukemic cells with chromosome 9–specific and chromosome 22–specific probes, each labeled with a different fluorescent marker, one is able to identify and compare the size of the normal chromosomes to those having undergone a translocation.

13. Metaphase chromosomes can be identified by size, shape, banding patterns, or hybridization to fluorescent probes (chromosome painting). Chromosome painting is a technique for visualizing chromosomes using fluorescent probes. In this method, DNA probes labeled with specific fluorescent tags are hybridized to metaphase chromosomes. After unbound probes are washed off, the chromosomes are visualized with fluorescence microscopy. Each chromosome then fluoresces with a different combination of fluorescent tags. Then a computer analyzes the tags and assigns a false color image. This way each chromosome can be identified by its false-color image and size.

14. Polytene chromosomes are present in larval salivary glands of the fruit fly *Drosophila melanogaster*, and are also present in cells in other dipteran insects and in plants. These enlarged interphase chromosomes, which can be observed with a light microscope, form as a result of multiple rounds of DNA replication (polytenization) without chromosome separation or cell division. Polytene chromosomes consist of multiple gene copies, which when transcribed provide the cells with an abundance of mRNA encoding proteins required for larval growth and development.

15. Replication origins are the points at which DNA synthesis is initiated. The centromere is the region to which the mitotic spindle attaches. The telomeres are specialized structures located at the ends of linear chromosomes. (a) The chro-

mosome would not be capable of being duplicated during S phase. (b) The chromosome could be replicated, but it may not be segregated evenly to the two daughter cells. The centromere is responsible for proper segregation of the duplicated chromosomes; without it the chromosomes will be distributed to the daughter cells by chance.

16. Because DNA polymerase is unable to initiate synthesis of a nucleotide strand, RNA primers must first be introduced to synthesize both leading and lagging strands. Eventually, the RNA region is degraded and DNA polymerase fills in the missing nucleotides. In the case of the lagging strand, DNA polymerase will fill in this gap using an upstream Okazaki fragment. At the very ends of the chromosome, there is no upstream Okazaki fragment, and thus it is not possible to replace the RNA nucleotides with DNA. Thus the chromosome shortens after each round of replication, this being known as the end replication problem. While telomeres do not halt this problem, they ensure that the lost DNA is non-coding DNA. Rather than lose protein or RNA encoding DNA, the genome instead loses non-coding telomere DNA.

ANALYZE THE DATA

1. a. If the kanamycin-resistance gene were localized solely in the chloroplast DNA of leaf cells of the spectinomycin-resistant plants, no kanamycin-resistant plants could be generated from the leaves since the kanamycin-resistance gene is under the control of a nuclear promoter. However, kanamycin-resistant offspring were obtained from leaves of the original spectinomycin-resistant transgenic plants. Thus, these data suggest that the kanamycin-resistance gene was transferred to the nucleus of the leaf cells that grew into kanamycin-resistant plants. A 3:1 ratio of kanamycin-resistant offspring to sensitive offspring among self-pollinating kanamycin-resistant plants indicates a Mendelian (i.e., nuclear) pattern of inheritance. A cross between two resistant plants in which a single copy of the kanamycin-resistance gene had been integrated into one autosomal chromosome would give a 3:1 ratio of kanamycin-resistant to sensitive offspring.

   b. Because no chloroplasts are inherited from the paternal (pollen-producing, transgenic plant), the only source of a kanamycin-resistance gene in the progeny seedlings was the haploid pollen nucleus from the transgenic, kanamycin- and spectinomycin-resistant parent. Each kanamycin-resistant seedling contained the kanamycin-resistance gene, as expected. But every one of the kanamycin-resistant progeny seedlings also contained the spectinomycin-resistance gene (seedlings 2, 3, 4, 7, 9). These findings suggest that the two antibiotic-resistance genes were transferred to the nucleus together in each case. Because only the spectinomycin-resistance gene was transcribed in chloroplasts and not the kanamycin-resistance gene, it is unlikely that the original mode of transfer of the kanamycin-resistance gene from a chloroplast to the nucleus was via an RNA intermediate. It is much more likely that a fragment of chloroplast DNA containing both the kanamycin- and spectinomycin-resistance genes was transferred from chloroplast DNA into the nuclear DNA of some of the leaf cells (i.e., the leaf cells that generated kanamycin-resistant plants).

c. Most of the original, spectinomycin-resistant plants had incorporated the spectinomycin- and kanamycin-resistance genes into their chloroplast DNA molecules, making them spectinomycin resistant. However, most of these plants did not integrate the kanamycin-resistance gene into one of their nuclear chromosomes. Only rare leaf cells selected for resistance to kanamycin had transferred the kanamycin-resistance gene into a nuclear chromosome. In the plants generated from these rare kanamycin-resistant leaf cells, kanamycin-resistance was inherited like any other nuclear autosomal gene, exhibiting Mendelian segregation patterns.

2. a. Slippage between the daughter and template strand during replication can result in an increase in repeats.

   b. An SDT1 promoter with 13 repeats results in the highest gene expression. Less than this results in lower expression as does more than 13 repeats. Q RT-PCR involves the extraction of mRNA from strains that contain varying promoters. The mRNA is then converted into cDNA, which is used as a template for a PCR reaction using SDT1 specific primers. If there is more gene expression, there will be more mRNA, which results in faster production of DNA in the Q PCR reaction.

   c. This experiment assumes that under selective pressure, it is possible to obtain yeast that are capable of altering the number of repeats in the promoter. This is because repeat number affects gene expression. In the absence of uracil, those cells that are capable of increasing uracil production will have a growth advantage, which is evident from this experiment. It is most likely that somewhere in the neighborhood of 13 repeats would allow the most uracil synthesis.

   d. One would conclude that histones do not associate with repetitive DNA in the promoter region. By decreasing the repeats, you would be more likely to see histones associated with those areas of the promoter.

7

TRANSCRIPTIONAL CONTROL OF GENE EXPRESSION

REVIEW THE CONCEPTS

1. In glucose media, lac repressor bound to the lac operator sterically blocks initiation by RNA polymerase: lac operon expression is repressed. After shifting to lactose media, lactose is transported into the cell where it binds the lac repressor. This causes the repressor to dissociate from the lac operator, allowing RNA polymerase to bind to the lac promoter and initiate transcription. In addition, the cell synthesizes cAMP because of the low concentration of glucose. cAMP binds to the catabolite activator protein (CAP), causing it to bind to the CAP site upstream from the lac promoter. The lac promoter is a weak promoter in the absence of CAP binding; CAP bound at the CAP site stimulates RNA polymerase binding and transcription initiation at the lac promoter, resulting in a high rate of transcription from the lac promoter and maximal induction of the lac operon.

2. In two-component regulatory systems, one protein acts as a sensor and the other protein is a response regulator. In *E. coli* the proteins NtrB (the sensor) and NtrC (the response regulator) regulate transcription of glutamine synthetase in response to the free glutamine concentration. NtrB has a sensor domain that binds glutamine and a His kinase transmitter domain with protein kinase activity. Under low glutamine conditions, NtrB undergoes a conformational change that activates the protein kinase activity and results in the transfer of a phosphate group to NtrC. The phosphorylated form of NtrC then induces transcription of glutamine synthetase.

3. RNA polymerase I is responsible for transcribing 18S and 28S rRNA genes. RNA polymerase II is responsible for mRNA transcription. RNA polymerase III transcribes

tRNAs, 5S rRNAs, and several other small RNAs. Since RNA polymerase II is uniquely inhibited by a low concentration of α-amanitin, one can determine if any gene requires this polymerase by measuring gene transcription in the presence and absence of this compound. If RNA polymerase II is responsible for transcribing a gene, then transcription should only occur in the absence of α-amanitin.

4. The CTD becomes phosphorylated by a subunit of TFIIH during transcriptional initiation. The CTD is then further phosphorylated by cyclin T-CDK9. CTD phosphates are removed when RNA polymerase II terminates transcription.

5. TATA boxes, initiators, and CpG islands are all promoter elements. The TATA box was the first to be identified because it is found in the promoters of most genes expressed at high level. These were the first genes to be subjected to in vitro transcription. Since the TATA box occurs at a fixed location relative to the transcription start site (~30 bases upstream), it was relatively easy to recognize in the DNA sequence.

6. To identify DNA-control elements within promoter regions, investigators utilize 5′ deletion mutants and linker scanning mutants. These mutants contain a loss of sequence in the promoter region. The wild-type DNA sequence and mutated DNA are separately introduced into cultured cells, or transgenic animals are generated with the wild-type and mutant DNA. Then the expression level of the associated gene or a fused reporter gene is assayed. When an activator-binding site is deleted, expression is reduced. When a repressor-binding site is deleted, expression is increased.

7. Promoter-proximal elements are located within ~200 base pairs of the transcription start site. Enhancers are located at greater distances, either upstream or downstream of the transcription start site. Enhancers continue to activate transcription when they are inverted or moved many kilobases from the transcription start site.

8. Once a putative control region is identified, DNA footprinting with a nuclear extract can identify the precise DNA sequence that is bound by a protein in the extract. This assay depends on the ability of protein factors to "protect" DNA from DNaseI digestion. The electrophoretic mobility shift assay can also be used to determine whether proteins in a cell extract bind specifically to a DNA sequence within the probe used. In this assay, an extract is incubated with a labeled fragment of DNA. If protein binds to the DNA, then it "shifts" in terms of its migration on a polyacrylamide gel. This technique is often used as an assay to purify DNA-binding proteins.

9. Transcriptional activators and repressors contain a modular structure in which one or more transcriptional activation or repression domains are connected to a sequence-specific DNA-binding domain, usually through a flexible domain.

10. The location of the sequence on the chromosome may influence whether it is expressed; for example, if the sequence is adjacent to a telomere, it will be in a

silent (heterochromatin) locus and will therefore not be transcribed due to condensation of the chromatin at that site. Second, DNA/histone methylation or deacetylation can lead to chromatin condensation and prevent transcription from that site. Second, DNA/histone methylation or deacetylation can lead to chromatin condensation and prevent transcription from that site.

11. CREB binding to its co-activator (CBP) is regulated by cAMP, which stimulates phosphorylation of CREB. The phosphorylated acidic activation domain within CREB is a random coil in the absence of CBP. However, in the presence of CBP the phosphorylated activation domain undergoes a conformational change to form two α helices that wrap around a larger globular domain of the co-activator. In contrast, nuclear receptors contain a larger activation domain that is regulated by the binding of a hydrophobic ligand. When ligand binds to these domains, they undergo a conformational change that generates a groove in the globular activation domain that binds a short α helix in a coactivator. Thus, while the phospho-CREB activation is a relatively short random coil until it interacts with a larger globular domain in a coactivator, nuclear receptors have a large, folded, inactive activation domain that undergoes a conformational change allowing it to bind a short α helix in a co-activator.

12. The first protein to bind to a RNA polymerase II promoter is the TATA box–binding protein (TBP), a subunit of TFIID. This protein folds into a saddle-like structure that binds to the minor groove of DNA near the TATA box and bends the DNA. TFIIB then binds and makes contact with both TBP and DNA on either side of the TATA box. TFIIF and Pol II bind and Pol II is positioned over the start site. TFIIE and TFIIH bind and a helicase activity unwinds the DNA generating an "open" complex with the template strand in the active site of the polymerase.

13. Integration of gene *X* near the telomere is not ideal for good expression of gene *X*. Telomeres are usually contained in heterochromatin, which is tightly packed and less accessible for the transcriptional machinery. If the yeast line used for expression contained mutations in the *H3* and *H4* histone genes, the outcome could be different, depending on the specific mutations. For example, if the DNA sequence encoding lysines in the histone N-termini were mutated so that glycine residues were substituted in their place, then repression of gene *X* would not take place. Repression would not take place in such a mutant because the glycine residues are not positively charged, similar to acetylated lysine residues. This prevents binding by SIR3 and SIR4, preventing the formation of heterochromatin. The resulting "open" chromatin structure at the telomere would facilitate RNA polymerase II and general transcription-factor binding, allowing gene expression.

14. A good prediction is that STICKY functions as a transcriptional repressor. Repressors contain two domains, one that binds DNA and a second that represses transcription. The bHLH domain is a DNA-binding domain that has been found in many different transcription factors. The Sin3-interacting domain is likely to associate with a Sin3 containing histone deacetylase complex. This complex can repress transcription because deacetylation of histones promotes a more closed chromatin conformation.

15.   A two-hybrid assay relies on unique yeast vectors, one referred as the "bait" and the other the "fish." The bait vector contains sequence encoding a DNA-binding domain (BD) and flexible linker region followed by a multiple cloning site (MCS). A cDNA encoding a known protein or protein domain is inserted into the MCS, in the proper frame to maintain the correct coding sequence. Four bait vectors need to be created for our experiment, one encoding the full length GR, and three others, each encompassing one of the modular domains in the protein. The bait vector also expresses a wild-type tryptophan (TRP) gene, allowing trp- yeast cells harboring this plasmid to propagate in media lacking TRP. A cDNA library generated from the pituitary cells is cloned into the MCS of multiple copies of another vector, which encodes sequence for a flexible linker and a strong activation domain (AD), needed to recruit co-activators and the transcription pre-initiation complex. This fish vector also contains a wild-type leucine (LEU) gene, allowing leu- cells containing the plasmid to grow in media lacking LEU. To identify a protein-protein interaction, the bait vector encoding the full-length GR and a library of fish vectors are transfected into engineered yeast cells containing a gene required for histidine (HIS) synthesis under the control of a UAS with binding sites for the bait vector's DNA-binding domain. Cells are plated on media lacking leucine and tryptophan to maintain the bait and fish vectors in cells, and lacking histidine to prevent cells from propagating unless the *HIS* gene is transcribed. *HIS* transcription requires that the DNA-binding domain of the bait hybrid bind the UAS, and the GR portion of the bait fusion protein has interacted with a fish protein fused in frame to the activation domain. Cells expressing both bait and fish plasmids survive and grow on media lacking *HIS* only if an interaction has occurred between GR and an interacting fusion fish protein. Recovery of the fish vector encoding the interacting protein and its subsequent sequencing reveals that protein X interacts with the GR. Having used the full-length GR to identify protein X, the experiment would be repeated with the three GR domain-encoding bait vectors. Instead, however, of using the entire cDNA library, each bait vector would be cotransfected into cells with the fish plasmid encoding protein X. Results would show that the GR ligand-binding domain is the one that binds protein X.

16.   Enhancers and promoter-proximal elements.

17.   a. Always be expressed
       b. Never be expressed
       c. Never be expressed
       d. Always be expressed

18.   *Similarities:*
       Both employ protein-binding regulatory DNA sequences/control elements and specific proteins that bind to these regulatory sequences to control gene transcription.

       Differences:
       Eukaryotes can alter gene transcription by adjusting the level of chromatin condensation.

Regulatory proteins are able to influence gene transcription by binding to regulatory sequences proximal or distal to the transcription start site in eukaryotes. There are three RNA polymerases in eukaryotes, one in prokaryotes.

19. Use recombinant DNA technology to make constructs with short fragments of gene X (including upstream and downstream sequences) adjacent to a reporter gene coding sequence such as green fluorescent protein (GFP). These constructs can be used to make transgenic organisms that can then be analyzed for GFP expression to identify the sequences capable of activating reporter gene expression.

20. The RNA-binding protein Tat binds to the RNA copy of a sequence called TAR, which then binds cyclin T. Consequently, the CDK9-cyclin T complex becomes activated, phosphorylates its substrates, and facilitates polymerase-mediated transcription elongation. In the absence of Tat, the transcripts are terminated prematurely. Antibodies against Tat would therefore prevent the infected cells from transcribing the viral genome because Tat would then be unable to bind the viral RNA.

21. No—low/medium level transcripts such as housekeeping genes often have a high frequency of CG sequences.

22. **Leucine zipper proteins:** contain the hydrophobic amino acid leucine at every seventh position in the sequence; bind to DNA as dimers.

    bHLH proteins: contain an N-terminal α helix with basic residues that interact with DNA; bind to DNA as dimers.

    Homeodomain proteins: conserved 60-residue DNA-binding motif similar to the helix-turn-helix motif of bacterial repressors; bind to DNA at positively charged (basic) residues that interact with phosphates in the DNA backbone as well as residues that interact with specific bases in the major groove of DNA.

    Zinc-finger proteins: 23- to 26-residue consensus sequence providing proteins with a region that folds around a central Zn^{2+} ion; found in DNA and non-DNA–binding proteins.

ANALYZE THE DATA

a. In untreated cells ($t = 0$ minutes), both wild-type and CARA cells transcribed equivalent amounts of ribosomal RNAs, suggesting that the mutant Pol I functions normally under control conditions. This finding is consistent with the observation that the growth rate of yeast harboring this mutation was indistinguishable from wild-type cells. However, when the drug rapamycin was added, Pol I transcription in wild-type cells was completely inhibited by 100 minutes after addition of the drug. In contrast, in the CARA cells, Pol I transcription was much less inhibited and continued at a significant level even at 100 minutes after addition of rapamycin. Consequently, Pol I transcription in the CARA cells was only partially inhibited by rapamycin.

b. As for rRNA precursor synthesis by Pol I analyzed in (a), both ribosomal protein mRNAs transcribed by Pol II and 5S rRNA synthesis by Pol III were rapidly inhibited by the addition of rapamycin in wild-type cells. Also, as observed for Pol I in (a), transcription of ribosomal protein and 5S rRNA genes was partially resistant to rapamycin and continued for a longer period after the addition of rapamycin in CARA cells compared to wild-type cells. Since the genetic modification of CARA cells directly affects Pol I, but not Pol II or Pol III, these differences from wild-type cells are probably a consequence of the continued Pol I transcription of the large rRNA precursor in CARA cells. Thus, these results indicate that transcription of ribosomal protein mRNAs by Pol II and of 5S rRNA by Pol III is coordinated with the supply of the large rRNA precursor. However, the molecular mechanisms that coordinate the transcription of these genes by Pol II and Pol III with the supply of the large rRNA precursor transcribed by Pol I remain to be discovered. This strategy for regulating the transcription of ribosomal protein genes and 5S rRNA genes seems logical. Since the ribosomal proteins and 5S rRNA are assembled onto the large rRNAs transcribed by Pol I during the synthesis of the ribosomal subunits, it makes sense for the cell to increase synthesis of these proteins and 5S rRNA in response to the supply of the large rRNAs.

c. In rich media, the CARA cells express the majority of genes, including the ribosomal protein genes, at a level similar to that of wild-type cells. Only a very small number of genes were expressed at more than twofold or less than one-half the level in wild-type cells. However, when the cells were shifted to poor media, the ribosomal mRNAs, but not other mRNAs, were expressed at a higher (twofold) level in the CARA cells than in the wild-type cells. Thus, these mRNAs were overrepresented in CARA cells under conditions that do not involve the use of drugs, but simply a decrease in nutrient supply. This overrepresentation in the mutant cells may be explained as follows: In wild-type cells, the shift to the poor media results in repression of Pol I synthesis of the large ribosomal RNAs and, coordinately, repression of Pol II synthesis of mRNAs encoding ribosomal proteins. In CARA cells, where the genetic manipulation affects Pol I and not Pol II, the continued expression of Pol I in low nutrient medium resulted in increased transcription of the ribosomal protein genes, confirming that transcription of ribosomal protein genes is stimulated in response to the production of the large rRNAs by Pol I.

8

POST-TRANSCRIPTIONAL
GENE CONTROL

REVIEW THE CONCEPTS

1. In the case of a protein-coding gene, gene control beyond regulation of transcriptional initiation can be regulated in several ways: 1) by controlling the stability of the corresponding mRNA in the cytoplasm; 2) by controlling the rate of translation; and 3) by controlling the cellular location so that newly synthesized protein is concentrated where it is needed.

2. True. Enzymes involved in mRNA capping, splicing, and polyadenylation are recruited to the phosphorylated CTD, which activates them. As only RNA polymerase II possesses a CTD, and RNA polymerase II is responsible for mRNA transcription, this ensures that these forms of processing only occur with mRNA.

3. These sequences are found near the intron/exon junctions, not the middle of the intron. Because of these sequences, the snRNPs of the spliceosome are recruited to the proper location on the mRNA. The role of the branch point A is to perform the first transesterification reaction, which eliminates the phosphodiester bond connecting the intron and the upstream exon. While RNA nucleotides have an OH group at both the 2' and 3' carbons, the 3' carbon of the branch point A is connected to an adjacent nucleotide. Thus, the OH group involved in this reaction must be at the 2' carbon.

4. The term hnRNA describes heterogeneous nuclear RNAs that consist of several different types of RNA molecules that are found in the nucleus. Small nuclear RNAs (snRNAs) bind to splice sites and participate in splicing reactions. Small nucleolar RNAs play a similar role in rRNA processing and can help to position

methyltransferases near methylation sites or convert uridines to pseudouridines. Micro RNAs (miRNAs) and short, interfering RNAs (siRNAs) are involved in gene silencing. Both are derived from longer precursor molecules and become part of the RISC complex.

5. Both types of splicing involve two transesterification reactions and similar intermediates and products. In group II intron self-splicing, the introns alone form a complex secondary structure involving numerous stem loops, whereas spliceosomal splicing utilizes snRNAs interacting with the 5′ and 3′ splice sites of pre-mRNAs, which form a three-dimensional RNA structure functionally analogous to the group II intron. Evidence that supports the idea that introns in pre-mRNAs evolved from group II self-splicing introns comes from experiments with mutated group II introns. In these experiments, domains I and V are deleted, and this yields a group II intron incapable of self-splicing. When RNA molecules equivalent to the deleted portions are added back "in trans" in the in vitro reaction, self-splicing is restored. This shows that portions of group II introns can be trans-acting like snRNAs.

6. In muscle cells, the internal polyadenylation site could be spliced out of the mature RNA when the fifth intron is removed. This would leave the site in the 3′ UTR as the sole polyadenylation site. In other cells, the fifth intron may not be spliced out. This would result in earlier polyadenylation and a shorter mRNA transcript. In this scenario, a muscle-specific splicing factor could facilitate removal of the fifth intron.

7. RNA editing is a type of pre-mRNA processing, altering the sequence of the pre-mRNA that results in a mature mRNA differing from the exons encoding it in genomic DNA. Although half of the sequence of some mRNAs may be altered in *Trypanosoma* and plant mitochondria and chloroplasts, only single-base changes have been observed in higher eukaryotes. A case for RNA editing in humans involves the serum protein apoB, which forms large lipoprotein complexes that carry lipids in serum. The apoB gene encodes two alternative forms of the protein, the ~240-kDa form (apoB-48) in intestinal epithelial cells and the ~500-kDa form (apoB-100), which is expressed in liver. RNA editing occurs in intestinal cells, where a single base alteration converts a codon for glutamine into a stop codon. The truncated protein is smaller and has a function distinct from the larger apoB-100 form, which as part of the low-density lipoprotein (LDL) particle, is responsible for transporting cholesterol to body tissues.

8. The nucleus is composed of a double membrane bilayer. These bilayers are amphipathic in nature, so it is extremely unfavorable for hydrophilic molecules to diffuse through. Instead, transport is restricted to the nuclear pores. The FG repeats are hydrophobic and line the entire nuclear pore complex. So while small molecules can freely pass through, larger hydrophilic molecules such as hnRNPs are blocked by these FG repeats, unless they are accompanied by nuclear export or import factors.

9. The mRNP exporter is a heterodimeric protein composed of the nuclear export factor 1 (NXF1) and the nuclear export transporter 1 (NXT1). NXF1 binds in

multiple places along mRNPs, together with other mRNP adapter proteins, including REF (RNA export factor) and SR proteins. Both exporter subunits interact with FG-nucleoporins, allowing them to move through the nuclear pore complex and into the cytosol. Protein kinases and phosphatases are thought to play a key role in the directional movement. In the nucleus, REF and SR proteins must be dephosphorylated in order to bind the mRNP exporter. In the cytoplasm, however, a kinase phosphorylates the adapter proteins, promoting the dissociation of the exporter from the mRNP. This dissociation results in a lower concentration of mRNP exporter-mRNP complexes in the cytoplasm than in the nucleus, allowing the complex to diffuse down its concentration gradient into the cytoplasm.

10. Short interfering RNAs (siRNA) can be synthesized to inhibit the function of any desired gene. siRNAs contain 21–23 nucleotides hybridized to each other so that two bases at each of the 3′ ends are single-stranded. The siRNA is introduced into cells, where it forms a complex with RISC. It then base-pairs with its target RNA and induces its cleavage, thereby eliminating the endogenous message. Cells containing siRNAs to TSC1 are likely to undergo uncontrolled cell growth because the loss of the TSC1 protein eliminates Rheb-GAP activity. This loss of activity causes abnormally high and unregulated levels of Rheb-GTP, which when bound to the mTOR complex would result in a high, unregulated activity of mTOR serving as an active kinase to phosphorylate a variety of substrates required to promote cell growth.

11. A plant deficient in Dicer activity shows increased susceptibility to RNA viruses because Dicer is not present to degrade a portion of the viral double-stranded intermediates that viruses synthesize during replication. Without Dicer, all of these viral mRNAs are available for further viral infection.

12. The deadenylation dependent decapping pathway involves the shortening of the poly(A) tail, which destabilizes the interaction between the 5′ cap and translation initiation factors. This results in removal of the 5′ cap, at which point the mRNA is degraded by nucleases from both the 5′ and 3′ ends. The other decapping pathway only involves 5′ degradation due to a 5′ cap that is more sensitive to decapping. Finally, it is possible for nucleases to degrade from within the mRNA and not at the 5′ or 3′ ends. P bodies are the sites of translation repression and/or degradation of mRNA. They are full of mRNA processing enzymes. If a decapping enzyme such as DCP1 had lower activity in a mutant cell, it likely would result in a build-up of mRNA molecules, increasing the size of the P bodies in the cell.

13. Proteins work at specific locations within the cell. Traditionally, it was thought that translation happens either in the cytoplasm or the ER, at which point the protein would be directed to its area of function. By localizing an mRNA, it is possible to synthesize proteins at specific places in the cell, thus eliminating the need to traffic this protein after it is translated. Ash1, a protein that inhibits mating type switching in yeast, has been shown to localize to the bud tip so that it is only present in the smaller daughter cells after budding. In the sea slug, neuronal mRNA were shown to localize to the synapse.

ANALYZE THE DATA

a.   Dicer is needed to cleave microRNAs from a larger transcript. The scientists tested the hypothesis that LAT encodes an miRNA by interfering with Dicer expression. Without Dicer, there should be no miRNA. If the LAT gene encodes an miRNA that protects cells from apoptosis, then cells that lack Dicer should not survive in the presence of the apoptosis-inducing drug. Indeed, in the presence of the LAT plasmid but in the absence of Dicer, the LAT gene did not protect cells from apoptosis. The observation that the Pst-Mlu fragment of the LAT gene provides protection from apoptosis suggests that an miRNA, if produced, would be derived from this region of the gene.

b.   The region of the LAT gene between the Sty-Sty sites is both necessary and sufficient to protect cells from apoptosis.

c.   Cells infected with either wild-type virus or rescued virus make RNAs of ~55 or 20 nucleotides containing sequence from the Sty-Sty region of the LAT gene. These RNAs are not observed in mock-infected cells or in cells infected with the DSty virus. The probe corresponding to the 3' end of the predicted stem can hybridize to RNAs containing the sequence of the 5' stem to which it is mostly complementary. It detects both the ~55- and 20-nucleotide RNAs. The probe corresponding to the 5' end of the predicted stem can hybridize to molecules containing the 3'-stem sequence to which it is complementary. This probe detects the ~55 nucleotide RNA. Thus, the ~55- nucleotide RNA contains both the 5'-stem sequence and the 3'-stem sequence, whereas the 20 nucleotide RNA contains only the 5'-stem sequence. These results strongly suggest that a 20-nucleotide miRNA is derived from the 5' end of an ~55-nucleotide pre-miRNA. The ~55-nucleotide RNA was probably processed from LAT RNA by Drosha in the nucleus. The 20-nucleotide RNA was probably processed by Dicer in the cytoplasm. The findings that a fragment of the LAT gene that includes the Sty-Sty region protects cells from apoptosis, that Dicer is needed to confer this protection, that part of the Sty-Sty region is predicted to form a stem-loop, a structure found in pre-miRNAs, and finally that this region gives rise to an ~55- and a 20-nucleotide RNA, the latter being the size of miRNAs, all strongly suggest that the LAT gene encodes an miRNA that protects cells from apoptosis.

d.   By forming an imperfect duplex with TGFβ mRNA, miR-LAT would inhibit translation of TGFβ. Accordingly, TGFβ expression would be reduced in cells infected with wild-type HSV-1. Because TGFβ expression potentiates cells to undergo apoptosis, a reduction in TGFβ would likely mean fewer apoptotic cells, as is the case when cells are infected with wild-type HSV-1. Thus, HSV-1, which encodes miR-LAT, can inhibit the cells it infects from undergoing apoptosis. Because no other HSV-1 genes are expressed in latently infected cells, the latently infected cells are not recognized by the immune system. Furthermore, because of the expression of LAT, they are more resistant to the induction of apoptosis than uninfected cells. Eventually, latency in some latently infected cells is overcome and those cells produce new progeny virus particles. These virus particles are transported down the axons of the virus-producing cells, where they escape from the neuron and infect surrounding cells in the region of skin enervated by that neuron, causing a recurrent cold sore.

9

CULTURING, VISUALIZING, AND PETURBING CELLS

REVIEW THE CONCEPTS

1. The major organelles of the eukaryotic cell are the nucleus, the endoplasmic reticulum, the Golgi complex, mitochondria, chloroplasts (plants), endosomes, and lysosomes (animals) or vacuoles (plants). The nucleus encloses the cell's DNA, contains the machinery for RNA synthesis, and physically separates the process of RNA synthesis from that of protein synthesis. The endoplasmic reticulum is responsible for the synthesis of lipids, membrane proteins, and secreted proteins. The Golgi complex processes and sorts secreted and membrane proteins. Mitochondria are the principal sites of ATP synthesis in aerobic cells. Chloroplasts are the site of photosynthesis, which produces ATP and carbohydrates in plant cells. Endosomes take up soluble material from the extracellular environment and lysosomes (in animal cells) degrade much of the material internalized in endosomes for use in biosynthetic reactions. In plant cells, vacuoles function like lysosomes in that they degrade material, but they also serve as storage sites for water, ions, and nutrients, and generate the turgor pressure that drives plant cell elongation. The cytosol is the unstructured aqueous phase of the cytoplasm excluding organelles, membranes, and insoluble cytoskeletal components. Cellular metabolism, protein synthesis, and signal transduction all take place in the cytosol.

2. Endocytosis occurs when a portion of the plasma membrane invaginates, forming a coated pit containing the protein or macromolecule. Numerous proteins, including clathrin, cover the cytosolic face of the pit and aid in its invagination and internalization.

Eventually, the pit pinches itself off into a membrane-bound vesicle, which is delivered to an early endosome where it gets sorted, either back to the plasma membrane, to the late endosome for further sorting, or to the lysosome where the contents are degraded. In the case of phagocytosis, large particles, including cells that have undergone apoptosis or bacteria, are enveloped by the plasma membrane and internalized for degradation in the lysosome.

3. The multiple membranes of mitochondria and chloroplasts act to create additional compartments with specialized functions within these organelles. The polarized stack of compartments that form the Golgi apparatus is associated with the assembly line organization of enzymes that modify many ER-derived products.

4. Specific types of cells in suspension may be isolated by a fluorescence-activated cell sorter (FACS) machine in which cells previously "tagged" with a fluorescent-labeled antibody are separated from cells not recognized by the antibody. The scientist selects an antibody specific for the cell type desired. Specific organelles are generally separated by centrifugation of lysed cells. A series of centrifugations of successive supernatant fractions at increasingly higher speeds and corresponding higher forces serves to separate cellular organelles from one another on the basis of size and mass (larger, heavier cell components pellet at lower speeds). This is often combined with density-gradient separations to purify specific organelles on the basis of their buoyant density.

5. Electron microscopy has better resolution than light microscopy, but many light-microscopy techniques allow observation and manipulation of living cells.

6. The total magnification of an image is described as the product of the magnification of the individual lenses, where the objective lens magnification immediately above the specimen is multiplied to that of the projection or eyepiece lens. Being able to clearly distinguish between two closely spaced points at even the highest total magnification is the ultimate goal because if the two objects are already blurred and cannot be discriminated at a lower magnification, simply increasing the magnification will have no effect. In fact, the formula defining the resolution (D) of a lens does not take magnification into account and is written as $D = 0.61\lambda$ $N \sin a$, where λ is the wavelength of light used to illuminate the specimen, N is the refractive index of the medium (usually air) between the front face of the objective lens and the specimen, and a is the half-angle of the cone of light entering the face of the objective lens. $N \sin a$ is often referred to as the lens' numerical aperture, which is physically stamped on the barrel of the objective lens. Since only three of the values can be altered to achieve the best resolution (the smallest D possible), one has to either decrease the wavelength of light or increase the numerical aperture by gathering more light into the front face of the objective lens. In most circumstances, therefore, the limitations include the use of wavelengths in the visible spectrum and the ability to gather more light to increase the numerical aperture. Increasing the numerical aperture is accomplished by placing a drop of oil or water, which have greater refractive indices (1.5 and 1.3, respectively) relative to that of air (1), between the specimen and the objective lens.

7. Chemical stains are required for visualizing cells and tissues with the basic light microscope because most cellular material does not absorb visible light and therefore cells are essentially invisible in a light microscope. The chemical stains that may be used to absorb light and thereby generate a visible image usually bind to a certain class of molecules rather than a specific molecule within that class. For example, certain stains may reveal where proteins are in a cell but not where a specific protein is located. This limitation can be overcome by fluorescence microscopy, in which a fluorescent molecule may be either directly or indirectly attached to a molecule of interest which is then viewed by an appropriately equipped microscope. Only light emitted by the sample will form an image, so the location of the fluorescence indicates the location of the molecule of interest. Confocal scanning microscopy and deconvolution microscopy build on the ability of fluorescence microscopy by using either optical (confocal scanning) or computational (deconvolution) techniques to remove out-of-focus fluorescence and thereby produce much sharper images. As a result, these techniques facilitate optical sectioning of thick specimens as opposed to physical sectioning and associated techniques that may alter the specimen.

8. FACS (see Figure 9-2), whereby labeled cells pass through a laser light beam and the fluorescent intensity of light emitted is measured, allowing the computer to assign each cell with an electric charge proportional to the fluorescence. Fibroblasts having twice the amount of DNA (G2 phase) compared to the normal diploid cells will emit more fluorescence and therefore have a different electric charge, which allows them to be separated and collected.

9. shRNAs under the control of tissue-specific promoters introduced as DNA constructs by viral vectors should, theoretically, continually express the siRNAs, thereby permanently knocking down the protein of interest. An obvious limitation includes knocking down an essential gene that leads to cell mortality.

10. Certain electron microscopy methods rely on the use of metal to coat the specimen. The metal coating acts as a replica of the specimen, and the replica rather than the specimen itself is viewed in the electron microscope. Methods that use this approach include metal shadowing, freeze fracturing, and freeze etching. Metal shadowing allows visualization of viruses, cytoskeletal fibers, and even individual proteins, while freeze fracturing and freeze etching allow visualization of membrane leaflets and internal cellular structures.

11. A cell strain is a lineage of cells originating from a primary culture taken from an organism. Since these cells are not transformed, they have a limited lifespan in culture. In contrast, a cell line is made of transformed cells and therefore these cells can divide indefinitely in culture. Such cells are said to be immortal. A clone results when a single cell is cultured and gives rise to genetically identical progeny cells.

12. These cultured cells allow one to conduct experiments to determine the roles of individual genes and proteins in the function of fat and muscle cells, for instance the metabolism of lipids in adipocytes and the contraction of muscle.

13. Normal B lymphocyte cells can produce a single type of antibody molecule. However, such cells have a finite lifespan in culture. Researchers use cell fusion of B lymphocytes and immortalized myeloma cells to create immortalized, antibody-secreting cells. Such cells, called hybridoma cells, retain characteristics of both parent cells, allowing for production of a single-type, or monoclonal, antibody.

ANALYZE THE DATA

1. a.

Organelle	Marker molecule	Enriched fraction (no.)
Lysosomes	Acid phosphatase	11
Peroxisomes	Catalase	7
Mitochondria	Cytochrome oxidase	3
Plasma membrane	Amino acid permease	15
Rough endoplasmic reticulum	Ribosomal RNA	5
Smooth endoplasmic reticulum	Cytidylyl transferase	12

   b. The rough endoplasmic reticulum is denser than the smooth endoplasmic reticulum because it is found in a gradient fraction with a higher sucrose concentration (more dense solution).

   c. The plasma membrane represents the least dense fraction because it has been fragmented into small pieces that reseal to form small vesicle-like structures.

   d. Addition of a detergent to the homogenate would eliminate the basis for equilibrium density-gradient centrifugation, as each organelle membrane would be solubilized by the detergent. Subjecting a detergent-treated homogenate to equilibrium density-gradient centrifugation would most likely produce a single peak at relatively low-percent sucrose that would contain all marker molecules.

2. a. siRNA was successful in worms C and D. RT-PCR shows no amplification (and hence, no mRNA) from one of the gene products.

   b. The set of bands near the middle represents unc18, and the set near the bottom of the gel represents GLUT4. GLUT4 was used as a positive control to ensure that GLUT4 expression is not affected by unc18 knock-down.

   c. The western blot would show bands (indicating presence of Unc18 protein) in protein samples from worms A, B, and E; no bands in samples from worms C and D.

d.

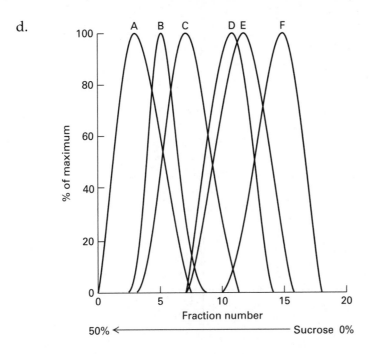

e. Colocalization can be demonstrated using confocal or fluorescence microscopy; a physical interaction can be demonstrated using fluorescence resonance energy transfer (FRET, see Figure 9-22).

10

BIOMEMBRANE STRUCTURE

REVIEW THE CONCEPTS

1. The spontaneous assembly of phospholipid molecules into a lipid bilayer creates a sheetlike structure that is two molecules thick. Each layer is arranged so that the polar head groups of the phospholipids are exposed to the aqueous environment on one side of the bilayer and the hydrocarbon tails associate with the tails of the other layer to create a hydrophobic core. In cross section, the bilayer structure thus consists of a hydrophobic core bordered by polar head groups. When stained with osmium tetroxide, which binds strongly to polar head groups, and viewed in cross section, the bilayer looks like a railroad track with a light center bounded on each side by a thin dark line.

2. The amphipathic nature of phospholipid molecules (a hydrophilic head and hydrophobic tail) allows these molecules to self-assemble into closed bilayer structures when in an aqueous environment. The phospholipid bilayer provides a barrier with selective permeability that restricts the movement of hydrophilic molecules and macromolecules across the bilayer. The different types of proteins present on the two faces of the bilayer contribute to the distinctive functions of each membrane, and control the movement of selected hydrophilic molecules and macromolecules across it.

3. The three main types of lipid molecules in biomembranes are phosphoglycerides, sphingolipids, and steroids. All are amphipathic molecules having a polar head group and a hydrophobic tail, but the three types differ in chemical structure, abundance and function.

4. Lipid bilayers are considered to be two-dimensional fluids because lipid molecules (and proteins if present) are able to rotate along their long axes and move laterally within each leaflet. Such movements are driven by thermal energy, and may be quantified by measuring fluorescence recovery after photobleaching, the FRAP technique. In this technique, specific membrane lipids or proteins are labeled with a fluorescent reagent, and then a laser is used to irreversibly bleach a small area of the membrane surface. The extent and rate at which fluorescence recovers in the bleached area, as fluorescent molecules diffuse back into the bleach zone and bleached molecules diffuse outward, can be measured. The extent of recovery is proportional to the fraction of labeled molecules that are mobile, and the rate of recovery is used to calculate a diffusion coefficient, which is a measure of the molecule's rate of diffusion within the bilayer. The degree of fluidity depends on factors such as temperature, the length and saturation of the fatty acid chain portion of phospholipids, and the presence/absence of specific lipids such as cholesterol.

5. Water-soluble substances are hydrophilic; they are therefore repelled by the hydrophobic core of the bilayer, which is composed of non-polar hydrocarbon tails of the phospholipids. Proteins that span the cell membrane (transmembrane proteins) provide a channel or passageway through which these substances can cross the membrane. The proteins fold such that their non-polar residues are in contact with the phospholipid bilayer and their polar residues line the channel through which the hydrophilic substances travel from one side of the cell membrane to the other.

6. Membrane-associated proteins may be classified as integral membrane proteins, lipid-anchored membrane proteins, or peripheral membrane proteins. Integral membrane proteins pass through the lipid bilayer and are therefore composed, of three domains: a cytosolic domain exposed on the cytosolic face of the bilayer; an exoplasmic domain exposed on the exoplasmic face of the bilayer; and a membrane-spanning domain, which passes through the bilayer and connects the cytosolic and exoplasmic domains. Lipid-anchored membrane proteins have one or more covalently attached lipid molecule, which embeds in one leaflet of the membrane and thereby anchors the protein to one face of the bilayer. Peripheral proteins associate with the lipid bilayer through interactions with either integral membrane proteins or with phospholipid heads on one face of the bilayer.

7. a. aquaporins
   b. porins

8. Cytosolic proteins are anchored to the plasma membrane by acylation or prenylation. In the case of acylation, an N-terminal glycine residue of a protein is covalently attached to the 14-carbon fatty acid myristate (myristoylation) or a cysteine residue in a protein is attached to the 16-carbon fatty acid palmitate (palmitoylation). Prenylation occurs when the –SH group on a cysteine residue at or near the C-terminus of the protein is bound through a thioether bond to either a farnesyl or a geranylgeranyl (prenyl) group. Cell-surface proteins and heavily glycosylated proteoglycans are present on the exoplasmic face of the membrane and are linked there by a glycophosphatidylinositol (GPI) anchor.

9. Since biomembranes form closed compartments, one face of the bilayer is automatically exposed to the interior of the compartment while the other is exposed to the exterior of the compartment. Each face therefore interacts with different environments and performs different functions. The different functions are in turn directly dependent on the specific molecular composition of each face. For example, different types of phospholipids and lipid-anchored membrane proteins are typically present on the two faces. In addition, different domains of integral proteins are exposed on each face of the bilayer. Finally, in the case of the plasma membrane, the lipids and proteins of the exoplasmic face are often modified with carbohydrates.

10. Detergents are amphipathic molecules. The hydrophobic part of a detergent molecule readily interacts with the hydrophobic tails of the phospholipids disrupting their interaction with each other; the hydrophilic part readily associates with water. This breaks up the organization of the lipid bilayer, ultimately leading to formation of micelle droplets, composed of a single phospholipid layer with the polar heads in contact with water and a hydrophobic core excluding water.

    Ionic detergents, like all detergents, bind to both the hydrophilic and hydrophobic regions of membrane proteins that have been exposed after lipid bilayer disruption. Because of their charge, they can also disrupt the ionic and hydrogen bonds holding together the secondary and tertiary structure of a protein and are thus useful for completely denaturing a protein.

    Non-ionic detergents do not denature proteins and are therefore useful for extracting membrane proteins while maintaining their native conformation. At concentrations below the critical micelle concentration, they also prevent the hydrophobic regions of proteins that have been extracted from the cell membrane from interacting with each other and forming insoluble aggregates.

11. a. peripheral

    b. lipid anchored

    c. integral membrane protein. No, a strong ionic detergent like SDS will denature the protein.

12. Lipid raft

13. a. Membrane phospholipids are synthesized at the interface between the cytosolic leaflet of the endoplasmic reticulum (ER) and the cytosol. Water-soluble, small molecules are synthesized and activated in the cytosol. Membrane-bound enzymes of the ER then link these small molecules to create larger, hydrophobic membrane phospholipids.

    b. Membrane phospholipids can be flipped from the cytosolic leaflet of the ER membrane to the exoplasmic leaflet. This process, mediated by flippases, results in the incorporation of newly synthesized phospholipids into both leaflets.

    c. Phospholipids can be moved from their site of synthesis to other membranes (e.g., to the plasma membrane). Some of this transport is by vesicles. Some is due to direct contact between membranes. Small, soluble lipid-transfer proteins also mediate transfer. The mechanism of phospholipid transfer between membranes is not yet well understood.

14. The common fatty-acid chains in phosphoglycerides include myristate, palmitdate, stearate, oleate, linoleate, and arachidonate (see Table 2-4). These fatty acids differ in carbon atom number by multiples of 2 because they are elongated by the addition of 2 carbon units. For example, the acetyl group of acetyl CoA is a 2-carbon moiety.

15. Fatty acids have very low solubility inside an aqueous-rich intracellular environment. Therefore, they associate with fatty-acid binding proteins (FABPs), which are cytosolic proteins that contain a hydrophobic pocket or barrel, lined by β sheets. This pocket provides a haven for the long-chain fatty acid, where it interacts in a noncovalent fashion with the FABP.

16. The key regulated enzyme in cholesterol biosynthesis is HMG (β-hydroxy-β-methylglutaryl)-CoA reductase. This enzyme catalyzes the rate-controlling step in cholesterol biosynthesis. The enzyme is subject to negative feedback regulation by cholesterol. In fact, the cholesterol biosynthetic pathway was the first biosynthetic pathway shown to exhibit this type of end-product regulation. As the cellular cholesterol level rises, the need to synthesize additional cholesterol goes down. The expression and enzymatic activity of HMG-CoA reductase is suppressed. HMG-CoA reductase has eight transmembrane segments and, of these, five compose the sterol-sensing domain. Sterol sensing by this domain triggers the rapid, ubiquitin-dependent proteasomal degradation of HMG-CoA reductase. Homologous domains are found in other proteins such as SCAP (SREBP cleavage activating protein) and Niemann-Pick C1 (NPC1) protein, which take part in cholesterol transport and regulation.

17. Most phospholipids and cholesterol membrane-to-membrane transport in cells is not by Golgi-mediated vesicular transport. One line of evidence for this is the effect of chemical and mutational inhibition of the classical secretory pathway. Either fails to prevent cholesterol or phospholipid transport between membranes, although they do disrupt the transport of proteins and Golgi-derived sphingolipids. Membrane lipids produced in the ER cannot move to the mitochondria by classic secretory transport vesicles. No vesicles budding from the ER have been found to fuse with mitochondria. Other mechanisms are thought to exist. However, presently these are poorly defined. They include direct membrane-membrane contact and small, soluble lipid-transfer proteins.

18. Statins block the conversion β-hydroxy-β-methylglutaryl linked to CoA (HMG-CoA) to mevalonate (an important intermediate in cholesterol synthesis) by competitively binding the enzyme necessary for this conversion (HMG-CoA reductase).

ANALYZE THE DATA

1. a. In the liposomes, there are no constraints on the diffusion of GFP-XR in the plane of the bilayer. Accordingly, when GFP-XR in a small zone of the liposome is bleached, diffusion of the bleached molecules out of, and of unbleached molecules from, the surrounding region into the bleached zone

will result in recovery of the fluorescence in that zone. In the cell, however, no recovery is observed, and these observations suggest that, in vivo, the lateral diffusion of GFP-XR is constrained.

b. XR in the cell membrane is not immobile as is the XR adhered to the microscope slide. Therefore, the lack of fluorescence recovery observed for XR in the cells is not the result of XRs being immobilized but rather the result of XRs being confined within a small domain.

c. In the cells, FRET is observed between GFP-XR and YFP-XR, as indicated by the data showing that yellow light is emitted when GFP is excited. Accordingly, these data suggest that XR molecules are clustered together. XR appears not to be randomly distributed in the plasma membrane as would be expected if the protein were free to diffuse. In liposomes, FRET has not occurred between XR molecules, as no yellow fluorescence is detected. Therefore, XR is not clustered in liposomes and is free to diffuse, as was also suggested by the FRAP and single particle tracking data above.

2. iv

11

TRANSMEMBRANE TRANSPORT OF IONS AND SMALL MOLECULES

REVIEW THE CONCEPTS

1. Like O_2 and CO_2, NO passively diffuses through membranes. As it is produced by an enzyme and accumulates in the endothelial cell cytosol, NO passively diffuses down its concentration gradient though the endothelial cell plasma membrane out of the cell and then passively diffuses through the plasma membrane into the cytoplasm of the smooth muscle cell, where it acts to decrease contraction.

2. Of the two at neutral pH, ethanol is the much more membrane permeant. It has no acidic or basic group and is uncharged at a pH of 7.0. The carboxyl group of acetic acid is predominantly dissociated at this pH and hence acetic acid exists predominantly as the negatively charged acetate anion. It is nonpermeant. At a pH of 1.0, ethanol remains uncharged and membrane permeant. The carboxyl group of acetic acid is now predominantly nondissociated and uncharged. Hence, acetic acid is now membrane permeant. Any difference in permeability is very small.

3. Uniporters are slower than channels because they mediate a more complicated process. The transported substrate both binds to the uniporter and elicits a conformational change in the transporter. A uniporter transports one substrate molecule at a time. In contrast, channel proteins form a protein-lined passageway through which multiple water molecules or ions move simultaneously, single file, at a rapid rate. The major contributor to the free-energy-driving transport through a uniporter is the entropy (ΔS) increase as a molecule moves from a high concentration to a low concentration.

4. The three classes of transporters are uniporters, symporters, and antiporters. Both symporters and antiporters are capable of moving organic molecules against an electrochemical gradient by coupling an energetically unfavorable movement to the energetically favorable movement of a small inorganic ion. The ΔG for bicarbonate has two terms, a concentration term and an electrical term, because bicarbonate is an anion. Glucose is neutrally charged and hence its ΔG for transport has only a concentration term. Unlike pumps, neither symporters nor antiporters hydrolyze ATP or any other molecule during transport. Hence, these cotransporters are better referred to as examples of secondary active transporters rather than as actual active transporters. The term active transporter is restricted to the ATP pumps where ATP is hydrolyzed in the transport process.

5. To be transported, a molecule must fit into the aquaporin channel and form hydrogen bonds with N-H groups of amino acids lining the channel. Although H^+ is smaller than H_2O, it cannot form the required hydrogen bonds. Glycerol is much larger than H_2O, but the three-carbon chain is flexible and the three OH groups can form the required hydrogen bonds.

6. Uniporters mediate substrate-specific facilitated diffusion. Uptake of glucose by GLUT1 exhibits Michaelis-Menten kinetics and approaches saturation as the concentration of glucose is increased. For any given substrate, GLUT1 displays a characteristic K_m at which concentration GLUT1 is transporting the substrate at 50% of V_{max}.

    a. Determination of the rate of erythrocyte (GLUT1) transport of substrate versus concentration allows the determination of K_m for glucose versus galactose versus mannose. The K_m for glucose will be lowest, indicating that GLUT1 is glucose-specific, not galactose- or mannose-specific.

    b. Despite its smaller size, ribose cannot bind to GLUT1 as glucose does because it cannot form the same noncovalent bonds—and thus cannot be transported.

    c. Using Equation 11-1, we find that at 5 mM, GLUT1-expressing cells transport glucose at 77% maximal rate, whereas at 2.8 mM, the cells transport at 65% maximal rate.

    d. Liver cells convert glucose to glycogen, which maximizes the glucose gradient across the plasma membrane.

    e. Tumor cells often express a higher number of glucose transporters than normal cells.

    f. When insulin is low, GLUT4 is stored in intracellular vesicles. Insulin induces a rapid increase in the V_{max} for glucose uptake by stimulating fusion of these vesicles with the plasma membrane, thereby increasing the number of plasma membrane glucose transporters.

7. The four classes of ATP-powered pumps are: P-class, V-class, F-class, and ABC superfamily. Only the ABC superfamily members transport small organic molecules. All other classes pump cations or protons. The initial discovery of ABC superfamily pumps came from the discovery of multidrug resistance to chemotherapy and the realization that ultimately this was due to transport proteins (i.e., ABC superfamily pumps). Today, the natural substrates of ABC superfamily pumps are thought to be small phospholipids, cholesterol, and other small molecules.

8. Direct hydrolysis of the phosphoanhydride bond would result in release of the bond energy as heat, which would thus be "lost." By first transferring the phosphate bond to an aspartate (D) residue, the P-class ATPase uses the released bond energy to drive a conformational change in the protein from the E1 to the E2 state.

9. A rise in cytosolic Ca^{2+} concentration causes activation of calmodulin. Some Ca^{2+}-ATPase pumps are activated by Ca^{2+}-calmodulin, which lowers the cytosolic Ca^{2+} concentration by pumping Ca^{2+} either into the sarcoplasmic reticulum/endoplasmic reticulum or out of the cell. An anti-calmodulin drug would inhibit this negative feedback mechanism, leaving higher Ca^{2+} concentration in the cytosol for a longer period of time. In skeletal muscle cells, the result would be to prolong the length and/or strength of muscle contraction.

10. These drugs irreversibly inhibit the H^+/K^+ ATPase in the apical membrane of stomach parietal cells. Although the inhibition of a given H^+/K^+ ATPase is irreversible, the cells eventually make more of the pump.

11. Membrane potential refers to the voltage gradient across a biological membrane. The generation of this voltage gradient involves three fundamental elements: a membrane to separate charge, a Na^+/K^+ ATPase to achieve charge separation across the membrane, and nongated K^+ channels to selectively conduct current. The major ionic movement across the plasma membrane is that of K^+ from inside to outside the cell. Movement of K^+ outward, powered by the K^+ concentration gradient generated by Na^+/K^+ ATPase, leaves an excess of negative charges on the inside and creates an excess of positive charges on the outside of the membrane. Thus, an inside-negative membrane potential is generated. These potassium channels are referred to as resting K^+ channels. This is because these channels, although they alternate between an open and closed state, are not affected by membrane potential or by small signaling molecules. Their opening and closing are nonregulated; hence, the channels are called nongated. K^+ channels achieve selectivity for K^+, versus, say, Na^+, through coordination of the nonhydrated ion with carbonyl groups carried by amino acids within the channel protein. The ion enters the channel as a hydrated ion, the water of hydration is exchanged for interaction with carbonyl residues within the channel, and then as the ion exits the channel it is rehydrated. Within the confines of the channel protein structure, Na^+, unlike K^+, is too small to replace fully the interactions of water with those with amino acid–carried carbonyl groups. Because of this, the energetic situation is highly unfavorable for Na^+ versus K^+.

12. Expression of a channel protein in a normally nonexpressing cell permits the patch clamp assessment of channel properties. Typically, the cell used is a frog oocyte. Frog oocytes do not normally express plasma membrane channel proteins. Channel protein expression may be induced by microinjection of in vitro–transcribed mRNA encoding the protein. Frog oocytes are large and hence technically easier to inject and to patch clamp than other cells. One can then vary the ionic composition of the medium and determine whether the presence of Na^+ or of K^+ supports ionic movement through the channel.

13. Plant cells, unlike animal cells, are surrounded by a cell wall. This cell wall is relatively stiff and rigid. The hyperosmotic situation within the plant vacuole that typically constitutes most of the volume of the plant cell is resisted by the rigid cell wall and the cell does not burst. Overall, a plant cell is considered to have a turgor pressure because of the hyperosmotic vacuole. The Na^+/K^+ ATPase is key to animal cells avoiding osmotic lysis. Animal cells have a slow inward leakage of ions. In the absence of a countervailing export, this would result in osmotic lysis of the cells even under isotonic conditions. The main countervailing export is the net transport of cations by Na^+/K^+ ATPase (3 Na^+ ions out for 2 K^+ in).

14. The six oxygens in the main-chain carbonyl or side-chain carboxyl groups that bind each of the two Na^+ ions in the transporter are exquisitely positioned with a geometry similar to that of the water molecules with which Na^+ associates in solution. At one site, the carboxyl group of the bound leucine provides one of the coordinating oxygens. When Na^+ ions bind to the oxygens, they lose their water of hydration. The increase in entropy that occurs when hydration water molecules are freed promotes Na^+ ions binding at both sites. K^+ ions (and water molecules themselves) are too big to bind the six oxygens in the proper geometry and so do not compete with Na^+.

15. Glucose uptake from the intestinal lumen into the epithelial cells is driven by symport with 2 Na^+ ions by a 2 Na^+/glucose symporter. Binding of two Na^+ ions and one glucose molecule to high-affinity, outward-facing sites in the protein causes a series of conformational changes in the symporter that eventually allows Na^+ and glucose to be released from low-affinity sites facing the cytosol. Transport by this symporter is energetically favorable because movement of Na^+ ions into the cell is driven by both its concentration gradient and the transmembrane voltage gradient. Transport of two Na^+ ions into the cell provides ~6 kcal of energy—enough to generate an intracellular glucose concentration that is 30,000 times higher than in the intestinal lumen.

16. The Na^+/K^+ ATPase located on the basolateral surface of intestinal epithelial cells uses energy from ATP to establish Na^+ and K^+ ion gradients across the intestinal epithelial cell plasma membrane. Cotransporters couple the energetically unfavorable movement of glucose and amino acids into epithelial cells to the energetically favorable movement of Na^+ into these cells. The accumulation of glucose and amino acids here is an important example of secondary active transport. Tight junctions are essential for the process because they seal the interstitial space between cells and hence allow the transport proteins in the apical and basolateral membranes of the epithelial cell to be effective. Effective transport could not be achieved through a leaky cell layer. The coordinated transport of glucose and Na^+ ions across the intestinal epithelium creates a transepithelial osmotic gradient. This forces the movement of water from the intestinal lumen across the cell layer and hence promotes water absorption from sport drinks.

ANALYZE THE DATA

a. Transepithelial transport of glucose requires the cotransport of Na^+ down its concentration gradient to drive glucose against its concentration gradient from the apical medium into the cells. This coupling cannot occur when the concentration of Na^+ in the apical medium is low, as is the case for curve 2. Under these conditions, there would be no inward gradient of Na^+ across the apical membrane to drive glucose uptake. Subsequently, for the glucose to move from the cells into the basolateral medium, the glucose is transported down its concentration gradient and thus does not depend on Na^+ and its concentration in the basolateral medium. Accordingly, the concentration of Na^+ in the basolateral medium is not a factor.

b. In order for the epithelial cells to maintain a low Na^+ concentration, the Na^+ must be continually pumped out of the cells by the Na^+/K^+ ATPase. Ouabain inhibits this pump and thus would prevent the cells from keeping their cytosolic $[Na^+]$ low. If Na^+ cannot be pumped out, then the Na^+ gradient driving glucose transport into the cells would dissipate, and glucose transport into the cells would cease. The observation that glucose transport ceases only when ouabain is added to the basolateral medium suggests that the Na^+/K^+ ATPase is localized at and/or only functions in this membrane domain.

c. The drug inhibits the Na^+-glucose symporter.

12

CELLULAR ENERGETICS

REVIEW THE CONCEPTS

1. The pmf is generated by a voltage and chemical (proton) gradient across the inner membrane of mitochondria and the thylakoid membrane of chloroplasts. Like ATP, the pmf is a form of stored energy, and the energy stored in the pmf may be converted to ATP by the action of ATP synthase. Proton diffusion will decrease the proton gradient across the mitochondrial inner membrane, thus decreasing the pmf and therefore the formation of ATP from ADP, leaving the cell with less energy to carry out its energy-dependent processes. At high doses, DNP can be fatal.

2. The unique properties of the mitochondrial inner membrane include the presence of membrane invaginations (termed cristae), a higher than normal protein concentration, and an abundance of the lipid cardiolipin. The cristae increase the surface area of the inner membrane, thereby increasing the total amount of membrane and hence electron-transport chain components: ATP synthase molecules and transporters of reagents and products of the citric acid cycle and oxidative phosphorylation are all increased. The higher concentration of proteins involved in electron transport and ATP synthesis further increases the capacity of mitochondria to synthesize ATP. Finally, cardiolipin enhances the barrier properties of the inner membrane by reducing the membrane's permeability to protons.

3. Glycolysis does not require oxygen, but the citric acid cycle and the electron-transport chain do require oxygen to function. In the case of the citric acid cycle, oxygen is not directly involved in any reaction, but the cycle will come to a halt as NAD^+ and FAD levels drop in the absence of oxygen. For electron transport, oxygen is required as an

electron acceptor. In the absence of oxygen, certain eukaryotic organisms (facultative anaerobes) as well as certain cells (mammalian skeletal muscles during prolonged contraction) can produce limited amounts of ATP by glycolysis (a process known as fermentation).

4. In glycolysis, NAD^+ is reduced to NADH. When oxygen is present, the electrons of this electron carrier are eventually donated to the electron-transport chain in oxidative phosphorylation. In the absence of oxygen, oxidative phosphorylation does not occur, and thus NADH is not oxidized, eventually leading to a shortage of NAD^+. Fermentation reactions oxidize NADH, replenish the store of NAD^+, and thus allow glycolysis to continue.

5. Electrons are passed to the electron carrier NADH in the cytoplasm. The NADH must then travel to the inner mitochondrial membrane, where the electrons are utilized by the electron-transport chain. NADH molecules can freely pass through the outer mitochondrial membrane through the channel protein, porin, which allows free passage of small molecules. As the mitochondrial inner membrane is impermeable to NADH, electrons cannot be passed directly from this electron carrier to the electron-transport chain. Instead, electron shuttles, such as the malate-aspartate shuttle, indirectly transfer electrons via intermediates that shuttle back and forth across the inner membrane.

6. Fatty acids are oxidized in the mitochondria and the persoxisome, but unlike the mitochondria, oxidation in the peroxisome does not generate ATP. Oxidation of very long chain fatty acids in peroxisomes leads to their degradation. In the case of the human genetic disease X-linked adrenoleukodystrophy (ALD), however, this oxidation is defective because the ATP-binding cassette (ABC) ABCD1 transporter, localized to peroxisome membranes, is unable to import very long chain fatty acids into this organelle. ALD patients have elevated levels of these fatty acids in their plasma and tissues, which is somehow associated with the degeneration of the adrenal gland and the myelin sheath surrounding nerve fibers in the brain.

7. Prosthetic groups are small nonpeptide organic molecules or metal ions that are tightly associated with a protein or protein complex. Several types of heme, an iron-containing prosthetic group, are associated with the cytochromes. The various cytochromes in the electron-transport chain contain heme prosthetic groups with different axial ligands, and as a result, each cytochrome has a different reduction potential so that electrons can move only in sequential order through the electron carriers.

8. The multiprotein complexes in the electron-transport chain pass electrons between proteins within a single complex and from one complex to the next. As this occurs, the electrons undergo a drop in electrical potential. The released energy is used to transport protons across the inner membrane, generating a proton gradient that is later used by ATP synthase. Respiration supercomplexes would be an efficient way to quickly pass electrons from one complex to the next, increasing the speed at which the process could occur. They have been demonstrated using native PAGE and electron microscopy. CoQ functions within the

inner membrane, bringing electrons to complex III. In the process, it picks up protons on the matrix side of the inner membrane, later depositing them into the inner membrane space. This is only possible because CoQ is soluble within the membrane bilayer.

9. The underlying reason for the difference in ATP yield for electrons donated by $FADH_2$ and NADH is that the electrons carried in $FADH_2$ have less potential energy (43.4 kcal/mol) than the electrons carried in NADH (52.6 kcal/mol). Thus, $FADH_2$ transfers electrons to the respiratory chain at a later point than does NADH, resulting in the translocation of fewer protons, a smaller change in pH, and fewer synthesized ATP molecules.

10. ATP synthase is comprised of the F_0 and F_1 components. F_0 is embedded in the mitochondrial inner membrane, and it is through this component that protons travel from one side of the inner membrane to the other, as they move down their electrochemical gradient. F_1 is connected to F_0 and possesses the ATPase catalytic activity. One of its subunits is responsible for nucleotide binding, inter-acting with ADP and P_i molecules, and releasing ATP following its formation. This process is mirrored in vesicle acidification, although rather than generating ATP, the ATPase portion of the enzyme hydrolyzes ATP, which is coupled to proton pumping against a gradient. This build-up of protons results in a pH decrease in the vesicle lumen. The overall process can be thought of as ATP synthesis in reverse.

11. Aerobic bacteria carry out oxidative phosphorylation by the same processes that occur in mitochondria (and are simpler and easier to work with than mitochon-dria). Glycolysis and the citric acid cycle take place in the bacterial cell cytosol, while electron-transport components are localized to the bacterial plasma mem-brane. Since electron transport takes place at the plasma membrane, the pmf is generated across the plasma membrane. In addition to using the pmf to synthe-size ATP, aerobic bacteria also use the pmf to power uptake of certain nutrients and cell swimming.

12. In addition to providing energy to power ATP synthesis, the pmf also provides the energy used by several active transport proteins to move substrates into the mitochondria and products out of the mitochondria. The OH^- gradient, which results from generation of the pmf by electron transport, is used to move HPO_4^{2-} into the matrix, and the voltage gradient contribution of the pmf drives exchange of ADP for ATP.

13. The Q cycle functions to double the number of protons transported per electron pair moving through a specific complex of the electron-transport chain and thereby maximizes the pmf across a membrane. In mitochondria, the specific complex is the $CoQH_2$-cytochrome c reductase complex, while in chloroplasts it is the cytochrome bf complex, and in purple bacteria it is the cytochrome bc_1 com-plex. Using mitochondria as an example, the Q cycle is believed to function as follows: $CoQH_2$ arrives at the Q_o site on the intermembrane space side of the $CoQH_2$-cytochrome c reductase complex; it delivers two electrons to the com-plex, and releases two protons into the intermembrane space. Next, one electron

is transported directly to cytochrome c while the other partially reduces a CoQ molecule bound to the Q_i site on the inner side of the complex, forming a CoQ semiquinone anion. CoQ dissociates from the Q_o site and is replaced by another $CoQH_2$, which delivers two more electrons and releases two protons to the intermembrane space. As before, one electron is transferred to cytochrome c, but the other combines with the CoQ semiquinone anion at the Q_i site to produce $CoQH_2$, thus regenerating one $CoQH_2$. In sum, the net result of the Q cycle is that four protons are transported to the intermembrane space for every two electrons moving through the $CoQH_2$-cytochrome c reductase complex.

14. False. While ATP is generated in photosynthesis, this energy is used to create sugars, which the cells use in a variety of different processes (including respiration). The endosymbiont hypothesis explains that primitive prokaryotes that were capable of ATP generation or sugar production were engulfed by eukaryotic cells through phagocytosis, producing the double membrane seen in these organelles. The prokaryote and eukaryote developed a symbiotic relationship and eventually the prokaryote lost its independence.

15. $6CO_2 + 6H_2O \longrightarrow 6O_2 + C_6H_{12}O_6$ O_2-generating photosynthesis uses the energy of absorbed light to create, via electron donation to quinone, the powerful oxidant P^+ form of the reaction center chlorophyll. This, in turn, acts to remove electrons from H_2O, a poor electron donor. The electrons are then passed along an electron-transport chain, and the stored energy is converted to other forms for subsequent use in ATP synthesis and carbon fixation. The O_2 is not used in subsequent reactions in this pathway and thus is a by-product of the removal of electrons from H_2O.

16. Photosynthesis consists of four stages. During stage 1, which occurs in the thylakoid membrane, light is absorbed by the reaction center chlorophyll, a charge separation is generated, and electrons are removed from water, forming oxygen. During stage 2, electrons are transported via carriers in the thylakoid membrane to the ultimate electron donor, $NADP^+$, reducing it to NADPH, and protons are pumped from the stroma into the thylakoid lumen, producing a proton gradient across the thylakoid membrane. During stage 3, protons move down their electrochemical gradient across the thylakoid membrane through F_0F_1 complexes and power ATP synthesis. Finally, during stage 4, the ATP and NADPH generated in the earlier stages are used to drive CO_2 fixation and carbohydrate synthesis. CO_2 fixation occurs in the stroma and carbohydrate (sucrose) synthesis occurs in the cytosol.

17. Chlorophyll a is present in both reaction centers and antenna. Additionally, antennas contain either chlorophyll b (vascular plants) or carotenoids (plants and photosynthetic bacteria). Antennas capture light energy and transmit it to the reaction center, where the primary reactions of photosynthesis occur. The primary evidence that these pigments are involved in photosynthesis is that the absorption spectrum of these pigments is similar to the action spectra of photosynthesis.

18. Photosynthesis in green and purple bacteria does not generate oxygen because these bacteria have only one photosystem, which cannot produce oxygen. These organisms still utilize photosynthesis to produce ATP by utilizing cyclic electron flow to produce a pmf (but no oxygen or reduced coenzymes), which can be utilized by F_0F_1 complexes. Alternatively, this photosystem can exhibit linear, noncyclic electron flow, which will generate both a pmf and NADH. For linear electron flow, hydrogen gas (H_2) or hydrogen sulfide (H_2S) rather than H_2O donates electrons, so no oxygen is formed.

19. PSI is driven by light of 700 nm or less and its primary function is to transfer electrons to the final electron acceptor, $NADP^+$. PSII is driven by light of 680 nm or less, and its primary function is to split water to yield electrons, as well as protons and oxygen. During linear electron flow, electrons move as follows: PSII (water split to produce electrons) $\longrightarrow$ plastoquinone (Q) $\longrightarrow$ cytochrome *bf* complex $\longrightarrow$ Plastocyanin $\longrightarrow$ PSI $\longrightarrow$ $NADP^+$. The energy stored as NADPH is used to fix CO_2 and ultimately synthesize carbohydrates.

20. The Calvin cycle reactions are inactivated in the dark to conserve ATP for the synthesis of other cell molecules. The mechanism of inactivation depends on the enzyme; examples include pH-dependent and Mg^{2+}-dependent enzyme regulation, as well as reversible reduction-oxidation of disulfide bonds within certain Calvin cycle enzymes.

21. Rubisco (ribulose 1,5-bisphosphate carboxylase) is a large enzyme present in the stromal space of the chloroplast. Rubisco is the enzyme responsible for adding (fixing) inorganic carbon in the form of CO_2 to the five-carbon sugar ribulose 1,5 bisphosphate, which is rapidly cleaved into two molecules of 3-phosphoglycerate that can be converted into starch and sugars.

ANALYZE THE DATA

a. The electron-transport system normally pumps protons out of the mitochondrial matrix, increasing the pH of the matrix; thus, the fluorescence of matrix-trapped BCECF would increase in intensity. The observed decrease in intensity of BCECF trapped inside the vesicles suggests that the vesicles have an inverted (inside-out) orientation, so that protons were pumped from the outside to the inside of the vesicles.

b. The concentrations of ADP, P_i, and oxygen should decrease over time as the process of oxidative phosphorylation utilizes oxygen as an electron acceptor and uses ADP and P_i to synthesize ATP.

c. Dinitrophenol compromises the pH gradient and the resulting equilibration of protons leads to an increase in the intravesicular pH and corresponding increase in emission intensity. Valinomycin, a potassium ionophore, affects the electric potential more than the pH gradient. Since BCECF fluorescence reflects the pH of the milieu, it is largely unaffected by valinomycin-induced changes in the transmembrane electric potential.

d. The fluorescence intensity inside the vesicles should remain constant over time since inner mitochondrial membranes from brown fat tissue would likely contain thermogenin, a protein that functions as an uncoupler of oxidative phosphorylation. Since thermogenin is a proton transporter, its presence would prevent the generation of a proton gradient and thus no fluorescence change would be expected.

e. To examine proton movement in chloroplast you would use the thylakoid membranes. In this case, the proton concentration is high in the thylakoid lumen and low outside, so the pH inside the thylakoid would increase as time passed, thus producing an increase in BCECF fluorescence.

13

MOVING PROTEINS INTO MEMBRANES AND ORGANELLES

REVIEW THE CONCEPTS

1. a. In the absence of ER membranes, the entire protein is translated and the ER signal sequence remains on the protein.

   b. When translation occurs in the presence of ER-containing microsomes, the protein is translated into the lumen of the microsomes. Following this process, the signal sequence is cleaved producing a smaller protein.

   c. Translation and translocation across the ER membrane are simultaneous processes. If they do not occur at the same, the protein is not properly imported into the ER where the signal sequence can be cleaved (although there are some examples of post-translational translocation).

2. a. The energy source for cotranslational translocation comes from the translation process itself—in other words, the nascent chain is pushed through the translocon channel. Please note, however, that as translation is completed a portion of the newly synthesized protein still resides within the translocon. This portion is drawn into the ER lumen rather than being pushed.

   b. In post-translational translocation, the newly synthesized polypeptide chain is drawn through the translocon by an energy input from ATP hydrolysis by BiP. BiP is luminal protein of the ER and is a member of the Hsc70 family of molecular chaperones. BiP-ATP activates by binding to the Sec63 complex that in turn binds to the Sec61 translocon complex. Activated BiP is enzymatically active and cleaves ATP to ADP plus P_i. It is BiP-ADP that binds to the entering, unfolded nascent chain. Sequential binding of BiP-ADP to the nascent chain serves to block any sliding of the chain back and forth in the translocon and to ratchet the nascent chain through the translocon.

c. Translocation into the mitochondrial matrix occurs through a bipartite Tom/Tim complex in which Tom is the outer membrane translocon and Tim is the inner membrane translocon. Three energy inputs are required. First, ATP hydrolysis by a cytosolic Hsc70 chaperone keeps the newly synthesized mitochondrial precursor protein unfolded in the cytosol. Second, ATP hydrolysis by multiple ATP-driven matrix Hsc70 chaperones may serve to pull the translocating protein into the matrix. Matrix Hsc70s interact with Tim44 and hence may be analogous to the BiP/Sec63 interaction at the ER membrane. Third, energy input from the H$^+$ electrochemical gradient or proton-motive force is required. The inside-negative membrane electric potential may serve to electrophorese the amphipathic matrix-targeting sequence toward the matrix.

3. SRP (signal recognition particle) acts as a cycling cytosolic factor for the translocation of ER targeted proteins. It binds to both the signal sequence and SRP receptor, a heterodimer associated with the ER membrane. In doing this, SRP initiates ribosome binding to ER membranes and positions the nascent chain proximal to the translocon. Both SRP and the SRP receptor are GTPases. The unfolded nascent chain then translocates. Cytosolic Hsc70 functions as a cytosolic factor required for protein translocation into mitochondria. It acts as a molecular chaperone to keep the post-translationally targeted mitochondrial precursor protein in an open, extended conformation. At least two different cytosolic proteins are required for translocation of peroxisomal matrix proteins. These are Pex5, the soluble receptor protein for matrix proteins containing a C-terminal PTS1 targeting sequence, and Pex7, the soluble receptor protein for matrix proteins containing an N-terminal PTS2 targeting sequence. A different receptor, Pex19, is required for peroxisomal membrane proteins.

4. Many membrane proteins are embedded in the membrane by virtue of transmembrane α-helical segment(s). Such segments can be referred to as topogenic sequences. These segments share general principles or properties. They tend to be about 20 amino acids long, a length sufficient to span the membrane, and hydrophobic, an appropriate property for a sequence embedded in the hydrophobic lipid bilayer. Application of these principles through computer algorithms is predictive. In brief, amino acid sequences of polypeptides may be scanned for hydrophobic segments of about 20 amino acids long. Each amino acid may be assigned a hydrophobic index value based on relative solubility in hydropbobic media versus water, and these values then can be summed by a computer for all 20 amino acid segments of a protein. Segments exceeding a threshold value are expected to be topogenic transmembrane segments. Internal signal anchor and stop-transfer anchor segments similarly can be identified. Such sequences alternate within a multipass membrane protein. Because of this, the overall arrangement of the protein can be predicted as described in detail in the text.

5. The UPR pathway up-regulates transcription of protein chaperones. It is thought that the timing of glycosylation modifications is one manner in which misfolded ER proteins are identified. Dislocation into the cytoplasm is necessary because the proteolytic machinery for these ER proteins is located in the cytoplasm.

6. The seven-sugar intermediate is synthesized by sugar addition to cytosolic-facing dolichol phosphate. The intermediate is flipped from the cytosolic face of the ER membrane to the luminal face. Further sugar additions then occur within the lumen of the ER. Short forms of the intermediate are on the wrong side of the membrane to add to nascent polypeptides within the ER lumen. Incomplete adductants within the ER lumen are located appropriately to N-glycosylate nascent polypeptide.

7. Several proteins facilitate the modfication or folding of secretory proteins within the ER. These include signal peptidase, BiP, oligosaccharyl transferase, various glycosidases, calnexin and calreticulun, protein disulfide isomerase, peptidyl-prolyl isomerase, and others. Of these, BiP and peptidyl-prolyl isomerase act to facilitate conformation changes. Protein disulfide isomerase facilitates the making/breaking of disulfide bonds to ensure correct protein folding. Calreticulin and calnexin are lectins that bind to glycoproteins during folding. The others all directly support the covalent modification of proteins within the ER lumen.

8. Each mutation has a different effect.
    a. Tom22 together with Tom20 act as outer mitochondrial membrane receptor proteins for N-terminal matrix targeting sequences. A defective Tom22 receptor protein would result in accumulation of mitochondrial matrix targeted proteins in the cytosol, possibly followed by their turnover within the cytosol.
    b. Tom70 signal receptor is an outer mitochondrial membrane protein recognizing multipass mitochondrial membrane proteins that have internal signal sequences. Mutation in Tom70 will have no immediate effect on mitochondrial matrix protein import, as Tom70 does not recognize this class of protein.
    c. Matrix Hsc70 has a role in the folding of matrix proteins. Also, it is one source of energy for powering translocation. Defective matrix Hsc70 should result in clogging the Tom/Tim translocon complex with incompletely translocated proteins.
    d. Retention of the matrix targeting N-terminal signal sequence because of a defective matrix signal peptidase might well result in defective folding of the imported protein. The sequence normally is removed.

9. On the whole, protein import into the mitochondrial matrix and the chloroplast stroma, topologically equivalent locations, is by functionally equivalent mechanisms. Functionally analogous proteins mediate each process. However, the proteins are not homologous, indicating a separate evolutionary origin of mitochondria and chloroplasts. Energetically, unlike the situation for mitochondria, there is no need for a membrane electrochemical gradient for import into chloroplasts. Presumably, stromal Hsc70 pulls proteins into the stroma.

10. This is basically a molecular ruler question. How many amino acids must span the Tom/Tim complex to expose the matrix-targeting sequence to the matrix-processing protease? DHFR in the presence of the drug methotrexate is locked into a folded state. A chimeric mitochondrial protein with folded DHFR fails to translocate fully into the mitochondria matrix. Instead, it is stuck in the Tom/Tim complex. The number of amino acids between the matrix targeting sequence

and the folded DHFR sequence could be varied to provide a molecular ruler. Any unfolded N-terminal DHFR sequence must be included within the ruler. With respect to channel length, an overestimate will result from this approach, as the matrix targeting sequence must be spaced out from Tom/Tim to be accessible for cleavage.

11. Catalase is responsible for breaking down H_2O_2 to H_2O. Catalase, like most other peroxisome-localized enzymes, contains a peroxisome-targeting sequence (PTS1) consisting of three amino acids, serine-lysine-leucine, at its C-terminus. This PTS1 is recognized and binds in the cytosol to the Pex5 receptor. The catalase-Pex5 heterodimer moves to the peroxisome membrane, where it interacts with the Pex14 receptor located on the membrane. In this position, the complex interacts with three membrane proteins—Pex2, Pex10, and Pex12—that facilitate the translocation of catalase into the peroxisome.

12. Separate mechanisms are used to import peroxisomal matrix and membrane proteins. Hence, mutations can selectively affect one or the other. Either can result in the loss of functional peroxisomes. One approach to determining whether the mutant is primarily defective in insertion/assembly of peroxisomal membrane proteins or matrix proteins is to use antibodies to ask by microscopy if either class of proteins localize to "peroxisomal" structures (e.g., peroxisome ghosts). An alternate approach is cell fractionation, in which the assay determines whether the appropriate proteins are present in a membrane organelle fraction.

13. The NLS. The nuclear import receptor binds to the NLS on the cargo molecule and brings it into the nucleus. Here, the receptor binds to Ran-GTP, causing release of the cargo. The receptor is thought to interact with the FG repeats that are commonly found on nuclear pore complex proteins, moving from one to the next as it passes through the nuclear pore.

14. Ran-guanine nucleotide–exchange factor (Ran-GEF) must be present in the nucleus and Ran-GAP must be in the cytoplasm for unidirectional transport of cargo proteins across the nuclear pore complex. When the Ran is bound to GTP, it has high affinity for cargo proteins. During nuclear export, Ran-GTP picks up cargo proteins in the nucleus and carries them to the cytoplasm. To be able to release the cargo on the cytoplasmic side, GTP must be hydrolyzed to GDP, and this process is stimulated by Ran-GAP. Once translocated back into the nucleus, Ran needs to be in the GTP-bound state to pick up more cargo. Ran-GEF in the nucleus stimulates the exchange of GDP for GTP, and this process of export can start again.

ANALYZE THE DATA

1. a. Messenger RNA lengths vary by steps of 20 codons or 20 amino acids each in corresponding synthesized product. When translated in the absence or presence of microsomes, only mRNAs of 130 and 150 codons produce a product that displays any difference in size with the addition of microsomes. The 130-codon mRNA gives a product that is either full length—showing no difference in size when compared to the minus microsome product—or a

smaller product that is the presumed result of signal peptidase cleavage. This suggests variable or incomplete accessibility of the product to signal peptidase. In contrast, the next-step-size-longer mRNA codes for product that is fully sensitive to signal peptidase cleavage when synthesized in the presence of microsomes. The key datum is the smaller size (faster mobility) of the product plus microsomes versus the product minus microsomes. Hence, the conclusion is that the prolactin chain must be somewhere between 130 and 150 amino acids in length for the signal sequence to be fully accessible for cleavage.

b. The polypeptide must be mostly α-helical. A 100-amino-acid polypeptide as an α-helix spans 150 Å, the length of the ribosome channel. Thirty amino acids span about 50 Å, the membrane thickness. In sum, a total length of about 160 amino acids (130-amino-acid spacer) is required to space the signal peptide out by 150 Å, the necessary minimum length. Only a 60-amino-acid spacer is required to give a minimally accessible signal peptide if the polypeptide were in extended conformation.

c. The fact that there is no prolactin cleavage if microsomal membranes are added after prolactin translation is complete indicates that prolactin must translocate cotranslationally.

d. A series of parallel reactions were done and a microsomal membrane fraction was prepared by centrifugation. No prolactin labeled polypeptide is seen in the membrane fraction for mRNA shorter than 90 codons. Hence, it is only at this nascent chain length that any engagement of ribosomes with SRP and hence binding to membrane occurs. If the mRNA is 110 codons or longer, the membrane fraction contains all labeled product found in the total reaction as indicated by the equal intensity of the gel bands. So, roughly a total chain length of 100 amino acids is required to expose the 30 amino acid signal peptide for SRP binding. The two bands seen with the 130-codon reaction are due to partial cleavage of the product by signal peptidase. The single band seen with the 150-codon reaction reflects full cleavage of the signal peptide by signal peptidase.

2. a. The fact that pro-insulin accumulates indicates that it has not folded properly because of the inability of PDI to arrange the correct disulfide bonding needed for processing to the Golgi and its eventual secretion as insulin. That PDI is present in equal amounts indicates that juniferdin has no effect on PDI synthesis; rather, it works to block the *activity* of PDI as an enzyme, which explains the pro-insulin accumulation.

b. A is Ire1, which would not change due to treatment, and because of the pro-insulin pile-up, it would lose its association with Bip, which is now bound to the improperly folded pro-insulin. Ire1 not bound to Bip dimerizes into an active endonuclease, which leads to the processing of the Hac1 mRNA and its translation into protein. Hence, increased the levels of Hac1 on blot B.

c. The increase in Hac1 leads to its translocation to the nucleus where it serves in the transcription of genes encoding proteins that will assist in folding and the progression of proteins through to the Golgi and their eventual secretion.

3. a. The smaller-sized gold particles are labeled Tim44 (i = inner) and the larger particles Tom40 (o = outer).

b. The N-terminus of ADH is a mitochondrial targeting protein and therefore it targets the actin bound to it to mitochondria. The actin signal in the cytosolic pool is the wild-type actin and you would expect to see it in this fraction. The fact that actin is the same size in each would indicate that it is processed in mitochondria. In this case, the N-terminus of ADH is cleaved off, so re-probing the blot with the N-terminus ADH antibody would produce a single (much smaller) band in the mitochondrial lane.

c. The shift is due to the fact that the N-terminal sequence responsible for targeting the proteins to mitochondria cannot be cleaved. Even though the sequence allows the proteins to bind to the receptors on the mitochondrial outer membrane, the cyanide has dissipated the proton-motive force needed to facilitate their movement into the mitochondria, where the protease resides to cleave the signal sequence. Furthermore, even if cells were supplemented with ATP, the lack of the proton-motive force would still prevent the translocation of the proteins into mitochondria.

14

VESICULAR TRAFFIC, SECRETION, AND ENDOCYTOSIS

REVIEW THE CONCEPTS

1. Palade and colleagues chose the pancreatic acinar cell because it is a specialized secretory cell that packages trypsinogen, chymotrypsinogen, and other digestive enzymes into secretory granules that then are released in response to signaling. Most protein synthesis in these cells is devoted to secretion. The vast majority of ribosomes are found in association with the rough endoplasmic reticulum. The cells are also decidedly polarized with a definite gradient in organelle distribution. In his seminal experiments, Palade took advantage of radioactive amino acids to label proteins, the majority of which were incorporated into molecules associated with membrane-bound ribosomes and, subsequently, with secretory organelles. After pulse-chase labeling and sectioning the cells for electron microscopy, the sections were covered with photographic emulsion and exposed. In this autoradiographic method, silver grains exposed to the radioactive decay from the newly labeled protein were reduced and, when the emulsion was developed, the appearance of the silver grains was compared under the electron microscope to their subcellular distribution. In a more contemporary method, cells can be transfected with a hybrid gene encoding two proteins. In this manner, the transcript encodes a membrane glycoprotein from vesicular stomatitis virus (VSV G) that is fused to green fluorescent protein (GFP), which is the tag for detection. Cells expressing this gene rapidly synthesize VSVG-GFP in the ER, which is then transported through the secretory pathway to the cell surface. Using fluorescence microscopy to detect GFP, the investigator can monitor the expression of the chimeric protein and its localization in live cells. Thus, both methods require that proteins be labeled in an early compartment so that their processing and transport can be followed over time. The second necessary requirement is to have a way to identify

the compartment containing the labeled proteins. In each case, a form of microscopy was used that complements the labeling reaction.

2. Coat proteins play two roles in vesicle budding: 1) They provide a scaffold that establishes membrane curvature; and 2) they interact with cargo proteins or cargo protein receptors to provide enrichment of certain proteins in the bud. Small GTPases of either the Sar or ARF family recruit coat proteins to membranes. It is the GTP-bound form of Sar or ARF that is active. The exact mechanisms by which each acts are unknown. The mechanism is particularly unclear in the case of ARF, which recruits clathrin and different adapter proteins at different sites within the cell. As noted above, vesicles are enriched in cargo proteins. Moreover, newly formed vesicles are programmed for subsequent fusion events by the selective inclusion of v-SNAREs and Rabs in their membrane. Dynamin is the one protein well known to have a role in pinching off vesicles at the cell surface and at the *trans*-Golgi network. In all known cases, these are clathrin-coated vesicles.

3. The observation that decoating Golgi membranes by treatment of cells with the drug brefeldin A (BFA) results in the redistribution of Golgi proteins to the ER suggests that COPI, the major Golgi-associated coat, has a role in stabilizing Golgi structure. To some extent, COPI may be equivalent to the dam that holds back the water in a reservoir. ARF1 is the small GTPase that recruits COPI to Golgi membranes. Mutation of ARF1 to give a GDP-restricted form of the protein would result in a GTPase that is now unable to recruit COPI to membranes. Since the COPI association with membranes is dynamic, this mutation would shortly lead to uncoating of Golgi membranes. Note that mutation of ARF1 to the GTP-restricted form would have the opposite effect; the ARF1 would now be permanently in the "on" state and the association of COPI with Golgi membranes would be permanently stabilized. Since GDP-restricted ARF1 produces the same phenotype as BFA, this suggests that the drug evokes a normal physiological possibility.

4. Coat proteins must assemble onto a membrane to form a vesicle and must disassemble to allow the vesicle to fuse with a target membrane. Accumulation of Golgi enzymes in transport vesicle indicates that EAGE allows COPI assembly and formation of vesicles but blocks the uncoating of COPI necessary for retrograde transport of Golgi enzymes for cisternal maturation. This inhibits cisternal maturation and anterograde transport of newly synthesized proteins from the ER to the plasma membrane.

5. Vesicle fusion involves two stages: first, a docking stage mediated by long coiled-coil proteins such as EEA1, and then a specific membrane fusion step mediated by SNAREs. The docking or tethering protein, EEA1, is recruited to vesicles by Rab proteins in their GTP-activated state. Rab5 plays a role in prompting vesicle fusion with early endosomes. Overexpression of the GTP-restricted form, the activated form, of Rab5 would have the effect of prompting such fusions and lead to enlarged early endosomes.

6. NSF, through its ATPase activity, catalyzes the dissociation of v-SNARE/ t-SNARE complexes. Such complexes are essential in specific membrane fusion at

several stages of the secretory and endocytic pathways. Why then does the Sec18 NSF mutation produce a class C phenotype accumulation of ER-to-Golgi transport vesicles? This can be explained readily if one considers the need for NSF to generate free v-SNAREs and t-SNAREs to support multiple rounds of vesicle membrane fusion. In the absence of NSF activity, vesicles bud from the ER but are unable to fuse with downstream membranes because of the lack of v-SNAREs. Vesicles accumulate at what is the first vesicle organelle fusion step within the secretory pathway.

7. Procollagen subunits, synthesized by ER-associated ribosomes and assembled in the ER, associate laterally into bundles in the Golgi. The procollagen bundles are too large to fit into the vesicles that go between Golgi compartments and have never been found in those vesicles. This observation supports the Golgi cisternal maturation model in which vesicles transport Golgi enzymes in the retrograde direction rather than transporting newly synthesized protein in the anterograde direction through the Golgi.

8. Lys-Asp-Glu-Leu (KDEL) and Lys-Lys-X-X (KKXX) are both retrieval sequences for ER proteins. KDEL is a sequence feature of soluble ER luminal proteins; KKXX is found on the cytosolic domain of ER membrane proteins. Retrieval of a normally ER luminal protein from the *cis*-Golgi is a COPI-dependent process. COPI is found on the cytosolic face of the *cis*-Golgi membrane. The KDEL-containing protein is within the lumen of the *cis*-Golgi. The two interact through a bridging membrane protein, the KDEL receptor. It is the KDEL receptor/KDEL-containing protein complex that is retrieved to the ER. In the cisternal progression model, *trans*-Golgi proteins, for example, must be retrieved to the *medial*-Golgi to generate a new *trans*-Golgi cisterna. This is a COPI-mediated process. There must be interactions between COPI and Golgi proteins to promote such retrieval.

9. There are four known clathrin adapter protein complexes: AP1 (*trans*-Golgi to endosome), GCA (*trans*-Golgi to endosome), AP2 (plasma membrane to endosomes), and AP3 (Golgi to lysosome, melanosome, or platelet vesicles). Each contains one copy of four different adapter protein subunits. The clathrin coat, unlike the COPI or COPII coat, is a double-layered coat with a core coat of adapter proteins and an external clathrin coat. Each adapter complex is different, but all are related. Presently, it is not known if the coat of AP3 vesicles contains clathrin. This is consistent with the possibility that evolutionarily the adapter complex may well be the core coat with clathrin an accessory layer added later.

10. I-cell disease is a particularly severe form of lysosomal storage disease. Multiple enzymes are lacking in the lysosome and the organelle becomes stuffed with nondegraded material and therefore generates a so-called inclusion body. I-cell disease is inherited; it is caused by a single gene defect in the *N*-acetylglucosamine phosphotransferase that is required for the formation of mannose 6-phosphate (M6P) residues on lysosomal enzymes in the *cis*-Golgi. This enzyme recognizes soluble lysosomal enzymes as a class and hence a defect in this protein affects the targeting of a large number of proteins. A defect in the phosphodiesterase that removes the GlcNAc group that initially covers the phosphate group on mannose 6-phosphate would also produce an I-cell disease phenotype.

Similarly, defects in mannose 6-phosphate receptors would affect the targeting of lysosomal enzymes as a class.

11. The *trans*-Golgi network (TGN) is the site of multiple sorting processes as proteins exit the Golgi complex. The sorting of soluble lysosomal enzymes occurs via binding to mannose 6-phosphate (M6P) receptors. Binding is pH dependent and occurs at the TGN pH of 6.5 but dissociates at the late endosomal pH of 5.0–5.5. Hence, lysosomal enzymes reversibly associate with M6P receptors. Clathrin/AP1 vesicles containing M6P receptors and bound lysosomal enzymes bud from the TGN, lose their coats, and subsequently fuse with late endosomes. Vesicles budding from late endosomes recycle the M6P receptors back to the TGN. Packaging of proteins such as insulin into regulated secretory granules is a very different process. This sorting is thought to be due to selective aggregation followed by budding. The TGN also may be the site of protein sorting to the apical and basolateral cell surfaces in polarized cells. This is the case in MDCK cells, a line of cultured epithelial cells, where there is direct basolateral-apical sorting at the TGN cells. In contrast, hepatocytes use different mechanisms for sorting to basolateral versus apical surfaces. Here, newly made apical and basolateral proteins are first transported in vesicles from the TGN to the basolateral surface and incorporated into the plasma membrane by exocytosis. From there, both basolateral and apical proteins are endocytosed in the same vesicles, but then their paths diverge. The endocytosed basolateral proteins are recycled back to the basolateral membrane. In contrast, the apically destined endocytosed proteins are sorted into transport vesicles that move across the cell and fuse with the apical membrane in a process termed transcytosis.

12. In infected MDCK cells influenza viruses bud only from the apical membrane, whereas vesicular stomatitis (VSV) viruses bud only from the basolateral membrane. Likewise, in the TGN, newly synthesized influenza HA coat protein is sorted into vesicles that fuse with the apical membrane and VSV G protein is sorted into different vesicles that fuse with the basolateral membrane. The virus coat proteins use the same processes by which cell plasma membrane proteins are sorted to the apical and basolateral membrane domains in polarized cells. Inhibition by a mimetic peptide but not by the mutant peptide indicates that the VSV G protein cytoplasmic domain contains a basolateral membrane targeting signal that most likely interacts with a component of the basolateral sorting and targeting mechanism, and that the tyrosine is part of that signal.

13. Within the endocytic pathway, there is a progressive acidification (increased hydrogen ion concentration) in compartments going from early to late endosomes to lysosomes. The pH drops from almost neutral to pH 4.5. The binding of LDL to LDL receptor is pH sensitive. At the cell surface, neutral pH, LDL binds to LDL receptor. At an acidic pH, pH 5.5, LDL dissociates from its receptor. LDL is then transported to lysosomes and LDL receptor is sorted and recycled back to the cell surface. Mannose 6-phosphate bearing lysosomal enzymes dissociates from mannose 6-phosphate receptors in the acidic pH late endosomal compartment. Elevating pH prevents this. Receptors become saturated with lysosomal enzymes and the cell no longer has the capacity to direct newly

synthesized lysosomal enzymes to lysosomes. Instead, the enzymes are secreted from the *trans*-Golgi network by constitutive secretion.

14. In terms of membrane topology, both the formation of multivesicular endosomes by budding into the interior of the endosome and the outward budding of HIV virus at the cell surface are equivalent. Important mechanistic features are shared. Both processes involve an ubiquitination step. In multivesicular endosome formation, cargo proteins to be included in the budding endosome and the Hrs protein are ubiquitinated. For HIV budding, it is the HIV Gag protein that is ubiquitinated. In closing off the budding endosome or the budding HIV a cellular ESCRT protein complex recognizes the ubiquitin, and cellular Vps4 is used later to dissociate the ESCRT complex. The viral Gag protein mimics the function of cellular Hrs, redirecting ESCRT complexes to the plasma membrane. ESCRT binds to the C-terminal portion of HIV Gag protein. One logical peptide inhibitor/competitor of HIV budding is a synthetic peptide corresponding to the portion of Gag protein that binds to ESCRT. Such a peptide might well compete or interfere with normal cellular proteins such as ESCRT binding to ubiquitinated Hrs.

15. Phagocytosis is the actin-mediated process used by some cells to engulf whole bacteria and other large particles. During the process, extensions of the plasma membrane envelop the ingested material, forming vesicles called phagosomes that are transported to the lysosome for degradation. Autophagy is the process whereby a double membrane organelle or autophagosome envelopes soluble cytosolic proteins, peroxisomes, or mitochondria and delivers them to the lumen of the lysosome for degradation. The three steps in the formation and fusion of autophagic vesicles are: 1) nucleation, whereby either a fragment of a membrane-bound organelle, probably the ER, forms a vesicle to randomly envelop a portion of the cytoplasm, or purposely to form around a particular organelle; 2) growth and completion, involving new membrane contributed to the autophagosome membrane, thereby facilitating it growth into a cup-shaped organelle; and 3) targeting and fusion of the intact sealed and double-membrane autophagosome and its contents to the lysosome.

16. Plasma membrane LDL receptors (LDLRs) bind LDL and transferrin receptors (TRs) bind ferrotransferrin from the extracellular environment at neutral pH. Once LDL has been internalized by receptor-mediated endocytosis, the low pH of the late endosome dissociates the LDL ligand from the LDLR, and the empty LDLR is recycled back to the plasma membrane while the LDL is digested in the lysosome. In contrast, the low pH of the late endosome dissociates the Fe^{3+} from the ferrotransferrin, leaving the apotransferrin form of the ligand bound to the TR. When the TR is recycled back to the plasma membrane, the neutral pH of the external environment dissociates the apotransferrin ligand from the TR.

17. The mutant LDLRs bind LDL normally but only randomly get internalized into clathrin-coated vesicles because the mutated cytoplasmic domains fail to interact with the AP2 complex. Inefficient RME of full-length LDLRs with a mutation in any one of the residues of the NPXY sequence in the LDLR cytoplasmic domain revealed that sequence functions as a RME sorting signal.

ANALYZE THE DATA

1. a. Only specific v-SNARE and t-SNARE combinations result in fusion. In this case, v-SNARE 1 can induce fusion only with membranes containing t-SNAREs of the plasma membrane, whereas v-SNARE 2 induces fusion only with membranes containing those of the Golgi. v-SNARE 3 appears to be a bit more promiscuous, permitting fusion with membranes containing either plasma membrane or vacuolar t-SNAREs, though it induces more rapid fusion with those of the vacuole.

    b. Because v-SNARE 1 induces fusion with the plasma membrane, this v-SNARE might be expected to be on vesicles that emerge from the trans-Golgi network. Alternatively, it might be present on endosomal membranes that cycle in and out of the plasma membrane. v-SNARE 2 fuses with the Golgi and thus might be expected to be on vesicles that mediate transport between the ER and Golgi or within the Golgi. Alternatively, v-SNARE 2 might be on vesicles moving retrogradely from the plasma membrane back to the Golgi. v-SNARE 3 might be on vesicles that move between the Golgi and the vacuole or between the vacuole and the plasma membrane.

    c. In yeast, mutations in a specific v-SNARE gene would be useful for determining the specific function of that SNARE. Temperature-sensitive mutations could be quite helpful, as incubation at the restrictive temperature may result in an increase in the number of unfused vesicles carrying cargo and these might be assessed to determine where in the secretory pathway these vesicles likely reside.

    d. The data show that the cytoplasmic domain of v-SNARE 2 competitively interferes with the ability of liposomes containing v-SNARE 2 to fuse with liposomes containing Golgi t-SNARES. Competition occurs only if the cytoplasmic domain is mixed with the liposomes containing the Golgi t-SNARES and not if it is incubated with liposomes containing v-SNARE 2. These findings indicate that the cytoplasmic domain of v-SNARE 2 binds to the Golgi t-SNARES and thereby interferes with the t-SNARES' ability to bind to intact v-SNARE 2. Yeast that overexpress the cytoplasmic domain of v-SNARE 2 would be predicted to exhibit a defect in the secretory pathway (similar to a dominant negative mutation in v-SNARE 2) at the fusion step in which this SNARE is needed.

2. a. It's everywhere because GFP is not a secreted protein so it wouldn't have a hydrophobic signal sequence to get it into the RER in the first place. The KDEL sequence would not be recognized because the protein itself is not in the right compartment.

    b. This would be expected for any resident ER protein; some of it might be lost as vesicles move to the Golgi. Luckily, these proteins have a KDEL sequence that is recognized by the KDEL receptor and the protein is shuttled back to the RER. Since PDI is an essential protein involved in catalyzing disulfide bonds between proteins, many proteins not having this modification would not fold properly and would therefore have abnormal or no function. If both alleles of a PDI gene were knocked out, the mouse would not survive, as PDI is an essential enzyme.

    c. Any antibody to a specific nuclear protein (e.g., histones).

3. a. The drawn cell should show fluorescence in the RER and nowhere else. The reason is that a) the protein is there as evident by Western blot and b) it is biologically active as evident by the assay above. The only explanation for the symptoms, then, is that the protein is not getting to the right place (i.e., the *cis*-Golgi), which is why there are no lysosomal enzymes modified with M6P and hence the I-cell symptoms.

    b. Repeat the immunofluorescence experiment. The fact that *N*-acetylglucosamine phosphotransferase was present at the RER would lead one to suspect that COPII proteins were missing; to confirm that this is the case, one could either stain for these proteins or look for their expression on Western blots.

    c. Transcytosis, where proteins (even apical bound ones) are first targeted to the basolateral membrane and then endocytosed and redirected to the apical membrane.

15

SIGNAL TRANSDUCTION AND G PROTEIN-COUPLED RECEPTORS

REVIEW THE CONCEPTS

1. Extracellular signals are made by signaling cells. Receptor proteins are present in target cells. Binding of extracellular signaling molecules to cell-surface receptors triggers a conformational change in the receptor, which in turn leads to intracellular signal-transduction pathways that ultimately modulate cellular metabolism, function, or gene expression. Intracellular signal transduction pathways are evolutionarily highly conserved.

2. Endocrine, paracrine, and autocrine signaling differ according to the distance over which the signaling molecule acts. In endocrine signaling, signaling molecules are released by a cell and act on target cells at a distance. In animals, the signaling molecule is carried to target cells by the blood or other extracellular fluids. In paracrine signaling, the signaling molecules are released and affect only target cells in close proximity. In autocrine signaling, the cell that releases the signaling molecule is also the target cell. Growth hormone is an example of endocrine signaling because the growth hormone is synthesized in the pituitary, located at the base of the brain, and travels to the liver via the blood.

3. The ligand-receptor complex that shows the lower K_d value has the higher affinity. Because the K_d for receptor 2 (10^{-9} M) is lower than that for receptor 1 (10^{-7} M), the ligand shows greater affinity for receptor 2 than for receptor 1. To calculate the fraction of receptors with bound ligand, $[RL]/R_T$, use Equation 15-2 $[RL]/R_T = 1/(1 + K_d/[L])$. For receptor 1, the K_d is 10^{-7} M and the concentration of free ligand $[L]$ is 10^{-8} M. Thus, the $[RL]/R_T$ for receptor 1 is 0.091, that is, only 9% of the receptors

have bound ligand at a free ligand concentration of 10^{-8} M. In contrast, the [RL]/R_T for receptor 2 is 0.91; 91% of the receptors have bound ligand.

4. To purify a receptor by affinity chromatography, a ligand for the receptor is chemically linked to beads used to form a column. A detergent-solubilized cell membrane extract containing the receptor is then passed through the column. The receptor will bind to the ligand attached to the beads and other proteins will wash out. The receptor can then be eluted from the column with an excess of ligand. Analyzing its molecular weight by SDS-PAGE may be sufficient to identify the recovered receptor or provide protein for sequencing.

    A similar column chromatography approach or an alternative affinity approach known as a pull-down assay can be used to isolate active $G_{\alpha s}$ protein. To isolate active $G_{\alpha s}$ with a pull-down assay, the adenylyl cyclase (AC) effector enzyme to which only the active form of $G_{\alpha s}$ binds is linked to beads either directly by chemical linkage or indirectly through an anti-AC antibody bound to protein A beads. Mixing the AC-coupled beads with a cell extract and then pelleting the beads by centrifugation pulls down only active $G_{\alpha s}$ because inactive $G_{\alpha s}$ (the GDP-bound form) fails to interact with the effector enzyme. The amount of $G_{\alpha s}$ pulled down can be quantified under different signaling conditions.

5. Binding of the extracellular signal or, in the case of rhodopsin, absorption of a photon by retinal, causes a conformational change in the orientation of the transmembrane helices that changes the conformation of the C3 loop and C4 segment (and in some cases the C2 loop) on the cytosolic side of the receptor. This increases the affinity of the receptor cytosolic region for binding and activating a G protein.

6. For trimeric G proteins in the inactive state, $G_{\alpha s}$ is bound to GDP and complexed with $G_{\beta \gamma}$. Upon ligand binding to its receptor, the receptor undergoes a conformational change that affects the associated trimeric G protein. GTP is exchanged for GDP and the $G_{\alpha s}$-GTP complex dissociates from the $G_{\beta \gamma}$ complex. The released $G_{\alpha s}$-GTP or $G_{\beta \gamma}$ complexes then activate downstream effector proteins. Hydrolysis of GTP to GDP by the intrinsic GTPase activity of $G_{\alpha s}$ returns $G_{\alpha s}$ to the inactive state bound to GDP and the $G_{\beta \gamma}$ complex. A mutant G_{α} subunit with increased GTPase activity would be expected to hydrolyze GTP to GDP at a faster rate and thus reduce the time that the G_{α} subunit remains in the active state. This, in turn, would lead to reduced activation of the effector protein.

7. As in Figure 15-18, the $G_{\alpha s}$ protein could be expressed as a CFP-fusion protein and adenylyl cyclase, instead of $G_{\beta \gamma}$, could be expressed as a YFP-fusion protein. In this case, association of $G_{\alpha s}$ with adenylyl cyclase would yield an increase in energy transfer and fluorescence at 527 nm when excited with 440 nm light.

8. Steps at which a single active component activates multiple targets amplify the signal. Each active receptor activates multiple $G_{\alpha s}$ proteins, and each active PKA initiates a short kinase cascade by phosphorylating multiple GPKs, each of which phosphorylates multiple glycogen phosphorylase enzymes to continue the signal amplification. In contrast, each active $G_{\alpha s}$ protein activates only one AC, and each cAMP participates in the activation of only one PKA, neither of which directly

amplifies the signal response. An increase in the number of receptors would have a greater effect on amplification because proportionally more $G_{\alpha s}$ proteins would be activated for the same level of epinephrine, whereas if only the number of $G_{\alpha s}$ proteins were increased, the signal response would still depend on the number of receptors activated.

9. Cholera toxin can penetrate the plasma membrane of cells. In the cytosol it catalyzes a chemical modification of G_{α} proteins that prevents hydrolysis of bound GTP to GDP. As a result, G_{α} remains in the active state. This causes continuous activation of adenylyl cyclase even in the absence of hormonal stimulation. The resulting excessive rise in intracellular cAMP leads to the loss of electrolytes and water into the intestinal lumen, producing the watery diarrhea and dramatic fluid loss characteristic of cholera infections.

10. Activation of muscarinic acetylcholine receptors in cardiac muscle slows the rate of heart muscle contraction. These receptors are coupled to an inhibitory G protein. Activation of this system causes a decrease in cAMP in the cell that leads to opening of K^+ channels on the cell membrane. The muscle cell becomes hyperpolarized, which reduces the frequency of muscle contraction. Rhodopsin is a G protein–coupled receptor that is activated by light. Rhodopsin contains a light absorbing pigment, 11-*cis*-retinal, that is covalently linked to opsin. In the presence of light, 11-*cis*-retinal is converted to all-*trans*-retinal. This activated opsin then interacts and activates transducin, an associated G protein. The activated G_{α}-GTP complex binds to the inhibitory subunit of a phosphodiesterase. The released catalytic subunits of the phosphodiesterase hydrolyzes cGMP to 5'-GMP. As a result, the cGMP level declines, leading to the closing of a nucleotide-gated ion channel. As with the cardiac muscle system, signal activation ultimately results in hyperpolarization of the photoreceptor cells.

11. Epinephrine binding to the β-adrenergic receptor causes an activation of adenylyl cyclase through the activation of $G_{\alpha s}$, a stimulatory G protein. In contrast, epinephrine binding to the α-adrenergic receptor causes an inhibition of adenylyl cyclase through the activation of $G_{\alpha I}$, an inhibitory G protein. An agonist acts like the normal hormone, which in this case would be epinephrine. Thus, agonist binding to a β-adrenergic receptor would result in activation of adenylyl cyclase. In contrast, an antagonist binds to the receptor but does not activate the receptor. Thus, antagonist binding to a β-adrenergic receptor would have no effect on adenylyl cyclase activity. In fact, it would reduce a normal epinephrine stimulated response because it would prevent epinephrine from binding to the receptor.

12. The epinephrine-cAMP signaling pathway—from binding of epinephrine to the receptor to activation of PKA—is essentially the same in all the cells. The downstream biochemical pathway activated is specified by the substrate(s) phosphorylated by PKA.

13. Receptor desensitization can involve phosphorylation of the receptor itself. The increase in cAMP levels as a result of ligand binding to the receptor leads to an activation of protein kinase A. Protein kinase A can phosphorylate target proteins as well as cytosolic serine and threonine residues in the receptor itself.

Phosphorylated receptor can bind ligand but is reduced in its ability to activate adenylyl cyclase. Thus, the receptor is desensitized to the effect of ligand binding. Phosphorylated receptors are resensitized by the removal of phosphates by phosphatases. A mutant receptor that lacked serine or threonine phosphorylation sites could be resistant to desensitization by phosphorylation and thus would continuously activate adenylyl cyclase in the presence of ligand.

14. AKAPs localize PKA activity to certain regions of cells to increase the speed of the signaling response. AKAP15 protein localizes PKA next to Ca^{2+} channels in heart muscle cells, which reduces the time that otherwise would be required for diffusion of PKA catalytic subunits from their sites of activation to the Ca^{2+}-channel substrates. A different AKAP in heart muscle anchors both PKA and cAMP phosphodiesterase (PDE) to the outer nuclear membrane. Localization of PKA at the nuclear membrane facilitates rapid entry of some of the activated PKA catalytic subunits into the nucleus, where they phosphorylate transcription factors. The proximity of the PDE provides for tight localized control of PKA activity through a negative feedback mechanism in which PKA phosphorylation of the PDE accelerates destruction of the cAMP thereby reactivating PKA inhibition

15. Cleavage of PIP_2 by phospholipase C generates IP_3 and DAG. IP_3 opens IP_3-gated Ca^{2+} channels in the endoplasmic reticulum (ER) membrane, resulting in release of Ca^{2+} from the ER. When ER stores of Ca^{2+} are depleted, the IP_3-gated Ca^{2+} channels bind to and open store-operated Ca^{2+} channels in the plasma membrane, allowing an influx of Ca^{2+}. To restore resting levels of cytosolic Ca^{2+} ($<10^{-7}$ M), Ca^{2+}-ATPase pumps located in the ER membrane and in the plasma membrane pump cytosolic Ca^{2+} back into the ER lumen and out of the cell. The principal function of DAG is to activate protein kinase C, which then phosphorylates specific target proteins.

16. Prior to stimulation of the IP_3-DAG signaling pathway, Ca^{2+}-ATPases establish a resting level of Ca^{2+} in the cytosol of $<10^{-7}$ M, at which level few calmodulin sites have Ca^{2+} bound and the calmodulin is inactive. IP_3 stimulates a rise in cytosolic Ca^{2+} concentration to $>10^{-5}$ M, at which level most calmodulin sites have Ca^{2+} bound and calmodulin is active. The Ca^{2+} binding affinity of calmodulin is exquisitely tuned to bind and release Ca^{2+} in response to the physiological changes of Ca^{2+} concentration in the cytoplasm used for signaling.

17. In rod cells, cGMP opens cation channels, whereas the primary activity of cGMP in smooth muscle cells, like cAMP, is to activate a kinase.

ANALYZE THE DATA

1. a. For the wild-type G protein, the activity of adenylyl cyclase is what you would expect. In the presence of GTP, there is a basal level of adenylyl cyclase activity, which can be greatly stimulated by the addition of isoproterenol. Isoproterenol binds to the β_2-adrenergic receptor and causes activation of adenylyl cyclase.

In comparing adenylyl cyclase activity in the presence of GTP or GTP-γS, again the expected result is seen. The addition of GTP-γS leads to an increase in adenylyl cyclase activity because GTP-γS is nonhydrolyzable. Thus, the $G_{\alpha s}$ subunit remains active, leading to prolonged activation of adenylyl cyclase. In the case of the mutant, again the addition of isoproterenol results in an increase in adenylyl cyclase activity as expected. The adenylyl cyclase activity, however, is not different in the presence of GTP or GTP-γS. Thus, the mutation causes an increase in the basal activity of adenylyl cyclase, likely due to a change in the GTPase activity.

b. In cells transfected with the mutant G protein, higher levels of adenylyl cyclase would be present relative to cells transfected with wild-type G protein. Thus, mutant-transfected cells would have higher cAMP levels, which would result in higher levels of active protein kinase A. The higher protein kinase A levels would result in more extensive phosphorylation of target proteins, which would affect normal cell development and proliferation.

c. From the GTPase results, it is clear that the mutation affects the intrinsic GTPase activity of the $G_{\alpha s}$ subunit. These results are consistent with the adenylyl cyclase results. For the mutant G protein, binding GTP or GTP-γS to the $G_{\alpha s}$ subunit leads to the same level of adenylyl cyclase activation because the $G_{\alpha s}$ subunit has greatly reduced ability to cleave GTP.

2. a. In this case, three of the four bars can be labeled however one likes because there is no FRET. The large peak is due to the PKA interacting with the β subunit, which is one of the two regulatory subunits of glycogen phosphorylase kinase; it is one of the subunits that is phosphorylated by active PKA. There is only one peak because the other subunits of glycogen phosphorylase do not bind to the catalytic domain of PKA.

b. The same combination—PKA plus β subunit—would produce emission at 535 nm because even though receptor activation is bypassed, increasing the intracellular cAMP levels would be sufficient to activate PKA.

c. If the α subunit could not be phosphorylated then fewer calcium ions would bind the δ subunit and thus glycogen phosphorylase kinase activity would not be maximally active. There will still be some activity because the β subunit is there and is able to be phosphorylated. Therefore, the graph should be increasing in glycogen phosphorylase kinase activity, but never reaching the levels seen under the wild-type condition.

3. a. cAMP concentration in normal intestinal epithelial cells rises and then falls, whereas in cells treated with cholera toxin, cAMP level rises and remains high. This is because, in cholera toxin-treated cells, $G_{\alpha s}$ is permanently in the GTP "on" state, which causes adenylyl cyclase to remain active, producing abundant cAMP.

b. Higher, because the abundant cAMP keeps PKA active, which subsequently leads to the activation of channels which indirectly causes water to follow the ions out of the cell.

16

SIGNALING PATHWAYS THAT CONTROL GENE EXPRESSION

REVIEW THE CONCEPTS

1. Both cytokine receptors and receptor tyrosine kinases (RTKs) form functional dimers upon binding their ligands. Upon dimerization, one of the poorly active cytosolic kinases phosphorylates the other on a particular tyrosine residue in the activation lip. This phosphorylation activates the kinase, which phosphorylates the second kinase in the dimer as well as other tyrosine residues in the receptor. Both cytokine receptors and RTKs then serve as docking sites for signaling molecules that bind to these phosphotyrosine sites. A major distinction between cytokine receptors and RTKs is that the RTK is itself the tyrosine kinase, whereas the cytokine receptor has no catalytic activity but rather is associated with a JAK kinase.

2. a. When GRB2 binds the Epo receptor, the Ras/MAP kinase signaling pathway is activated, resulting in the translocation of MAP kinase to the nucleus, where it phosphorylates and regulates the activity of transcription factors.

   b. When JAK phosphorylates and activates STAT5, STAT5 itself translocates to the nucleus as a homodimer and functions as a transcription factor. Epo signaling induces the production and survival of erythrocytes, or red blood cells (RBCs), which increases the oxygen carrying (aerobic) capacity of the blood. In most normal people and even elite athletes, RBCs comprise 40–50% in men and 35–45% in women of the volume of the blood (hematocrit). Epo doping can raise that level to 75%, which makes the blood very viscous and can cause heart failure and death.

3. A dominant-negative mutant of JAK cannot be phosphorylated and thus is inactive, but it binds normally to the Epo receptor blocking the binding of wild-type JAK.

4. GRB2 serves as an adapter molecule that contains both SH2 and SH3 domains. GRB2 binds to phosphotyrosine residues on activated receptor tyrosine kinases via its SH2 domain and binds to the guanine nucleotide exchange factor Sos via two SH3 domains. Sos then binds to and activates the Ras protein. Although all SH2 domains bind to phosphotyrosine residues, specificity is determined by the conformation of the binding pocket, which interacts with amino acids on the carboxy-terminal side of the phosphotyrosine.

5. Negative feedback occurs when a signaling pathway induces expression or activation of its own inhibitor. (a) When erythropoietin binds to its receptor, EpoR, an SH2 domain on the phosphatase SHP1 binds to a phosphotyrosine on the receptor. Binding induces a conformation change that activates SHP1, which is in close proximity to JAK (associated with EpoR). SHP1 dephosphorylates and inactivates JAK, thus inhibiting signal transduction. The erythropoietin signaling pathway also possesses negative feedback that triggers long-term downregulation in which STAT proteins induce the expression of SOCS proteins. SOCS proteins contain SH2 domains that bind EpoR, preventing the binding of signaling molecules. One SOCS protein also binds the activation lip of JAK2 and inhibits its kinase activity. SOCS proteins also recruit E3 ubiquitin ligases, which ubiquitinate JAKs and target them for degradation by the proteasome. (b) TGFβ signaling induces the expression of SnoN and Ski, two proteins that bind to Smads and inhibit their ability to regulate transcription. TGFβ signaling also induces the expression of I-Smads, which prevent the phosphorylation of R-Smads by the TGFβ receptor.

6. Constitutive activation is the alteration of a protein or signaling pathway such that it is functional or engaged even in the absence of an upstream activating event. For example, RasD is constitutively active because it cannot bind GAP and therefore remains in the GTP-bound, active state even when cells are not stimulated by growth factor to activate a receptor tyrosine kinase. Constitutively active Ras is cancer promoting because cells will proliferate in the absence of growth factors, and thus normal regulatory mechanisms for cell proliferation are bypassed.

    a. A mutation that resulted in Smad3 binding Smad4, entering the nucleus, and activating transcription independent of phosphorylation by the TGFβ receptor would render Smad3 constitutively active.

    b. A mutation that made MAPK active as a kinase and able to enter the nucleus without being phosphorylated by MEK would render MAPK constitutively active.

    c. A mutation that prevented NF-κB from binding to Iκ-B or that allowed NF-κB to enter the nucleus and regulate transcription even when bound to Iκ-B would render NF-κB constitutively active.

7. In the mating factor signaling pathway Ste7, another serine/threonine kinase is the substrate for Ste11. When the mating factor signaling pathway is activated, Ste7, Ste11, and the other relevant kinases in the cascade form a complex with the scaffold protein Ste5. Binding to Ste5 ensures that Ste7 is the only substrate to which Ste11 has access.

8. Maximal activation of protein kinase B requires 1) release of inhibition by its own PH domain, which is achieved when PI 3-phosphate binds the PH domain; 2) phosphorylation of a serine in the activation lip of protein kinase B by PDK1, which occurs when both protein kinase B and PDK1 are recruited to the cytosolic surface of the plasma membrane by binding PI 3-phosphates; and 3) phosphorylation of an additional serine residue located outside the activation lip. In muscle cells, insulin-stimulated activation of protein kinase B causes fusion of intracellular vesicles containing the GLUT4 glucose transporter with the plasma membrane, resulting in increased influx of glucose. Insulin-stimulated activation of protein kinase B also promotes glycogen synthesis because protein kinase B phosphorylates and inhibits glycogen synthase kinase 3 (GSK3), an inhibitor of glycogen synthase.

9. PTEN phosphatase removes the 3-phosphate from PI 3,4,5-triphosphate, thus reversing the reaction catalyzed by PI-3 kinase and rendering the PI phosphate unable to bind protein kinase B and PDK1. Loss-of-function mutations are cancer promoting because constitutive activation of protein kinase B results in constitutive phosphorylation and inactivation of proapoptotic proteins such as Bad and Forkhead-1. Cancers are typically characterized by cells that are resistant to apoptosis. A gain-of-function mutation in PTEN phosphatase would promote cell death by causing the apoptotic pathway to be active even in the presence of survival factors that signal through protein kinase B.

10. In multiple cell types, TGFβ activates a conserved signaling pathway that results in translocation of Smad2 or Smad3 to the nucleus in complexes with co-Smad4. Once in the nucleus, the Smads interact with other transcription factors to regulate the expression of target genes. The complement of these other transcription factors is cell-type specific, and thus the TGFβ signaling pathway will induce the transcription of different genes in different cell types.

11. TGFβ binds to its type II receptor either directly or when presented by the type III receptor. The type II receptor is a constitutively active serine/threonine kinase. When the type II receptor binds TGFβ, it forms a tetrameric complex consisting of two molecules of the type II receptor and two molecules of the type I receptor. The type II receptor then phosphorylates the type I receptor, activating the type I receptor as a serine/threonine kinase. The type I receptor then phosphorylates R-Smad2 or R-Smad3, inducing a conformational change that exposes a nuclear localization signal on the R-Smad. The R-Smad then forms a complex consisting of two molecules of R-Smad and one molecule each of co-Smad4 and importin-β. This complex translocates to the nucleus, where it interacts with other transcription factors to elicit changes in gene expression. A nuclear phosphatase continuously dephosphorylates and inactivates Smads in the nucleus.

Maintenance of Smad activity in the nucleus thus requires continued activation of Smads by TGFβ-activated receptors.

12. Hedgehog is covalently linked to cholesterol and also has a palmitoyl group added to the N-terminus. Cholesterol addition occurs during autoproteolytic cleavage of the protein into two fragments. Together, these modifications make the Hh signaling domain hydrophobic so that it remains tethered to the cell membrane. Tethering of Hh to cell membranes may limit its range of action in tissues, allowing spatial restriction of its action.

13. Hedgehog (Hh) protein binding to the Patched receptor relieves its inhibition of Smoothened delivery to the cell surface, thus initiating the Hh signaling pathway. Nonfunctional versions of either Hedgehog or Smoothened protein would block the Hh signaling pathway, yielding the same phenotype. In contrast, nonfunctional Patched would allow Smoothened delivery to the plasma membrane even in the absence of Hh protein, permanently activating the Hh signaling pathway.

14. When Hh activates the Patched receptor, Smoothened (Smo) is phosphorylated by β-adrenergic receptor kinase (BARK), which causes β-arrestin to bind to Smo and recruit the microtubule motor protein Kif3A. Kif3A moves Smo along the microtubules in the core of the cilium up the ciliary membrane. Concurrently, degradation of Gli to a repressor fragment is blocked, and the motor protein Kif7 moves Gli to the tip of the cilium where its transcription factor activity is turned on by an unknown mechanism. The dynein motor protein moves the activated Gli to the base of the cilium from where Gli moves into the nucleus to activate gene transcription. Mutations in the respective motor protein would inhibit each of these parts of the Hh signaling pathway and the pathway as a whole.

15. The signaling pathway that activates NF-κB is considered irreversible because I-κB, the inhibitor of NF-κB, is degraded when the pathway is activated. Thus, the signal cannot be switched off rapidly as with a kinase-induced signal that can be reversed by the action of an opposing phosphatase. The NF-κB pathway is eventually disengaged by negative feedback in which NF-κB stimulates the transcription of I-κB. However, synthesis of the I-κB inhibitor de novo is a relatively slow process, and thus, the NF-κB pathway can remain active for some time after the original stimulus, such as TNF-α, is removed.

16. Stimulation of a cell by an infectious agent or inflammatory cytokine activates β kinase to phosphorylate I-κBα. This stimulates polyubiquitination of I-κBα lysine 48 (K48) by an E3 ubiquitin ligase, which targets the I-κBα for degradation by a proteasome. The E3 ligase TRAF6 polyubiquitinates I-κBα lysine 63 (K63). Rather than targeting the protein for degradation, the K63-linked polyubiquitin chain acts as a scaffold for poly K63 ubiquitin-binding domain proteins to bring together the kinase TAK1 with its substrate IκBα.

17. Delta is a single-pass plasma membrane protein with an extracellular Notch-binding signaling domain. Unlike EGF precursors, which can be cleaved by proteases to release diffusible EGF signaling molecules, Delta remains intact for signaling and therefore can only activate Notch receptors on neighboring cells

with which it is in direct contact.

18. γ-secretase is an intramembrane protease. When it was found that γ-secretase cleaves the intramembrane domain of APP to yield the Aβ42 fragment that forms amyloid plaques, γ-secretase inhibitors were tested for treatment of Alzheimer's disease. Unfortunately, γ-secretase also generates the cytosolic fragment of Notch that translocates to the nucleus as part of that signaling pathway. Inhibition of γ-secretase activity in this signaling pathway produces unacceptable side effects.

ANALYZE THE DATA

1. a. This experiment reveals that MEK5 and MEKK2 co-localize within a complex because immunoprecipitation with a MEK5 antibody also precipitates MEKK2. However, these data give no evidence as to whether MEKK2 activates MEK5 or MEK5 activates MEKK2.

    b. MEKK2 is required for the activation of ERK. However, MEKK2 cannot activate ERK unless MEK5 is present. MEKK2 alone partially activates ERK5 because there is some endogenous MEK5, but in the presence of MEK5AA, which inhibits endogenous MEK5, MEKK2 cannot activate ERK5. These experiments clearly place ERK5 downstream of MEKK2 and MEK5 in the signaling pathway; however, they do not unambiguously order MEKK2 and MEK5. To do so would require co-expression of ERK5, MEK5AA, and a constitutively active form of MEKK2. If ERK5 were phosphorylated, then MEKK2 would be downstream of MEK5. If not, then MEKK2 would be upstream of MEK5.

2. a. Curve 2 in the Kss1 activation assay indicates that Ste7 is active even in the absence of Ste5. The similarity in the Fus3 curve 3 and the Kss1 curves 2 and 3 indicate that Ste7 can phosphorylate Fus3 and Kss1 approximately equivalently. Comparison of curves 2 and 3 for Fus3 and Kss1 indicates that Ste5 is required for Ste7 to activate Fus3 but is not required for Ste7 to activate Kss1. This supports the possibility that Ste5 plays more than a tethering role in the Fus3 pathway.

    b. As shown above, Ste7 phosphorylates Kss1 even in the absence of Ste5, so Kss1 is a fully active substrate. In contrast, Ste7 phosphorylates Fus3 poorly unless Ste5 is present, indicating that Fus3 alone is a poor substrate for Ste7 but its phosphorylation can be enhanced by Ste5. Mutation or replacement of the Fus3 MAPK insertion loop activates Fus3 as a substrate even in the absence of Ste5. The most logical conclusion is that in addition to tethering Fus3 and Ste7 together, interaction with Ste5 suppresses MAPK insertion loop inhibition of Fus3 phosphorylation. The inability of Fus3 to be phosphorylated by Ste7, except when it is scaffolded together by Ste5, would contribute to the segregation of the mating and starvation pathways.

17

CELL ORGANIZATION AND MOVEMENT I: MICROFILAMENTS

REVIEW THE CONCEPTS

1. Actin filaments (microfilaments) are composed of monomeric actin protein subunits assembled into a twisted, two-stranded polymer. Actin filaments provide structural support, particularly to the plasma membrane, and are important for certain types of cell motility. Microtubules are composed of α- and β-tubulin heterodimers assembled into a hollow, tubelike cylinder. Microtubules provide structural support, are involved in certain types of cell motility, and help generate cell polarity. Intermediate filaments are formed from a family of related proteins such as keratin or lamin. The subunits assemble to create a strong, ropelike polymer that, depending on the specific protein, may provide support for the nuclear membrane or for cell adhesion.

2. For actin filaments, polarity refers to the fact that one end is different from the other end. This difference is generated because all subunits in an actin filament point toward the same end of the filament. The end at which the ATP-binding cleft of the terminal actin subunits contact the next internal subunits is termed the (+) end because it is the preferred end for polymerization with the lower critical concentration. The opposite end, termed the (−) end, is the less preferred end for polymerization with the higher critical concentration. Polarity can be detected by electron microscopy in "decoration" experiments in which myosin S1 fragments (essentially myosin head domains) are incubated with actin filaments. The S1 fragments bind along the actin filament with a slight tilt, leaving the actin filament decorated with arrowheads that all point toward one end of the filament. The pointed end corresponds to the (−) end and the barbed end corresponds to the (+) end.

3. Cells utilize various actin cross-linking proteins to assemble actin filaments into organized bundles or networks. Whether the actin filaments form a bundle or a network depends on the specific actin cross-linking proteins involved and the structure of the actin cross-linking protein. Actin cross-linking proteins that generate bundles typically contain a pair of tandem (i.e., closely spaced) actin-binding domains, while actin cross-linking proteins that generate networks typically contain actin-binding sites that are spaced far apart at the ends of flexible arms.

4. Once actin has been purified, its ability to assemble into filaments can be monitored by viscometry, sedimentation, fluorescence spectroscopy, or fluorescence microscopy. The viscometry method measures the viscosity of an actin solution, which is low for unassembled actin subunits but increases as actin filaments form, grow longer, and become tangled. The sedimentation assay utilizes ultracentrifugation to pellet (sediment) actin filaments but not actin subunits, and thereby separates assembled actin from unassembled actin. Fluorescence spectroscopy measures a change in the fluorescence spectrum of fluorescent-tagged actin subunits as they assemble into actin filaments. Lastly, fluorescence microscopy can be used to visualize the assembly of fluorescent-tagged actin subunits into actin filaments. Viscosity; and fluorescence microscopy will distinguish between short and long filaments, sedimentation and florescence spectroscopy will not.

5. ATP–G-actin assembles onto the ends of actin filaments and the ATP bound to the subunit is subsequently hydrolyzed to ADP and phosphate. As a result, most of the filament consists of ADP–F-actin with the exception of a small amount of ATP–F-actin at the (+) end. At the (−) end, exposed ADP–F-actin dissociates to become ADP–G-actin. If a mutation in actin prevents the protein from binding ATP, actin filaments will fail to assemble (or may assemble at very low levels if the actin can still bind ADP). If a mutation prevents actin from hydrolyzing ATP, the subunits will assemble into a filament that will be unable to disassemble normally.

6. Treadmilling is a form of actin filament assembly that occurs when the rate of subunit addition at one end equals the rate of subunit loss at the other end so that filament length remains constant as subunits flow, or treadmill, through the filament. For treadmilling actin filaments, ATP-actin subunits add to the (+) end and ADP-actin subunits dissociate from the (−) end. Treadmilling occurs when the unassembled actin subunit concentration is greater than the C_c value for the (+) end but less than the C_c value for the (−) end, which is the condition at the steady state level of polymerization.

7. Inactivation of profilin should reduce the amount of actin polymer, may disrupt signaling pathways that involve this protein, and may allow spontaneous initiation of new actin filament assembly at random places in the cell. Inactivation of thymosin-β_4 should increase the cell's free actin concentration and consequently actin filaments. Inactivation of CapZ would cause actin filaments to grow uncontrollably at the (+) end. Inactivation of Arp2/3 will inhibit assembly of actin filaments and formation of branched networks.

8. CapZ blocks the actin filament (+) end, so growth would occur only at the pointed (−) end. Tropomodulin blocks the actin filament (−) end, so growth would

occur only at the barbed end. The profilin-actin complex allows addition to a free (+) end but not to a (−) end, so growth would occur only at the barbed end.

9. Both formin and WASp are self-inhibited by intramolecular interactions before activation, and both are activated by binding of their RBD regions with an activated Rho protein. Formin is activated by binding Rho-GTP, and WASp is activated by binding Cdc42-GTP. Both proteins also contain domains that recruit profilin-actin for polymerization. The formin FH2 domain dimerizes, creating a hinged ring structure that rocks to allow addition of actin subunits while maintaining connection to the growing filament's (+) end. The WASp acidic A domain activates the Arp2/3 complex to bind to the side of an existing actin filament and initiate the growth of a new filament onto a free (+) end.

10. All myosins use energy derived from ATP hydrolysis to "walk" along actin filaments. Depending on the specific type of myosin, this movement is used to generate contraction or to transport specific cellular components relative to actin filaments. All myosins are composed of one or two heavy chains (the motor subunit) and several light chains. The heavy chains of all types of myosin have similar head domains (which interact with actin, bind and hydrolyze ATP, and generate force to move) but different tail domains, which specify the particular cellular component that a given myosin recognizes and hence the function of each myosin. For each myosin type, different light chains may be present, but all associate with the neck region just adjacent to the head domain. Myosin II is the only myosin capable of forming bipolar filaments that pull actin filaments in opposite directions.

11. Myosin motility may be observed in the sliding filament assay. In this approach, myosin motors are adsorbed onto the surface of a glass coverslip and the coverslip is placed onto a slide to create a chamber. Fluorescent-labeled actin filaments and ATP are then delivered to the chamber and the myosin motors will walk along the actin filaments. Since the motors are attached to the coverslip, the movement of myosin causes the actin filaments to slide across the coverslip surface. ATP must be added to these assays to provide the energy for myosin movement. The sliding-filament assay can reveal the direction of myosin movement if the polarity of the moving actin filaments is known. The force generated by myosin can be measured with an optical trap, in which optical forces are used to determine the force just needed to hold a sliding actin filament still.

12. The principal contractile bundles of nonmuscle cells are the circumferential belt of epithelial cells, the stress fibers present on cells cultured on artificial substrates, and the contractile ring. The circumferential belt and stress fibers appear to function in cell adhesion rather than cell movement. The contractile ring generates the cleavage furrow during cytokinesis that eventually leads to division of a single cytoplasm into two. Only myosin II can produce contractile force in these contractile bundles.

13. Conformational changes in the myosin head couple ATP hydrolysis to movement. When ATP binds to myosin, it releases from actin and hydrolyzes ATP. Upon ATP hydrolysis, the myosin head undergoes a conformational change that

rotates the head with respect to the neck portion of the protein. ADP and P_i remain bound to the myosin head. This conformational change stores the energy released by ATP hydrolysis. In the next step, the myosin-ADP-P_i complex binds a new actin subunit followed by P_i release. When P_i is released, the myosin head pops back to its original conformation, moving the bound actin filament along with its own rotation. In this way, the chemical energy of ATP hydrolysis is coupled to the mechanical work of moving an actin filament.

14. Two main structural differences between myosin II and myosin V explain the differences in their properties and functions: 1) Myosin V proteins have a longer neck than myosin II, and 2) myosin II proteins can assemble into bipolar filaments involved in contractile functions. (Myosin V also has globular cargo binding domains at its tail.)

   The neck domain acts as a lever arm for the conformational change during the power stroke. Thus, the head of myosin V moves farther than the head of myosin II with each stroke. The ATPase activity is slower for myosin V than for myosin II. Thus, the head remains bound to actin during a longer portion of each cycle. The "duty cycle" is about 70%. This means that at any time during the cycle, one or both of the heads are bound to the actin filament, so the vesicle does not float away. As myosin V moves down a filament, its two heads follow a hand-over-hand movement. The longer duty cycle and large step size correspond with the function of myosin V, which is to provide the motor to move cargo along actin filaments.

   Bipolar complexes of myosin II work together in contraction. Hundreds of myosin heads may interact with an actin filament during muscle contraction. Each head remains bound only transiently, while other heads bind and move the filament. This arrangement allows for fast movement when loads are light, but it also allows the flexibility of greater force for heavier loads. The shorter duty cycle and cooperative action between many myosin II molecules fits the function of myosin II as a contractile protein.

15. The mechanism by which a rise in Ca^{2+} triggers contractions differs in skeletal and smooth muscle. In skeletal muscle, binding of Ca^{2+} to troponin C leads to muscle contraction. Contraction of smooth muscle is triggered by activation of myosin light-chain kinase by Ca^{2+}-calmodulin.

16. By stimulating PKA phosphorylation of MLCK, Albuterol inhibits the Ca^{2+}-calmodulin activation of MLCK, preventing phosphorylation of the regulatory light chain on smooth muscle myosin II that stimulates myosin activity for contraction.

17. Keratinocytes and fibroblasts have been used for experiments on cell locomotion. The movement of these cells is pictured as a series of four steps. First, the leading edge of the cell is extended by actin polymerization. Second, the newly extended membrane forms an attachment (containing embedded actin filaments) to the substrate, which anchors this portion of the cell to the substrate and prevents its retraction. Third, the bulk of the cell body is translocated forward, perhaps by myosin-dependent contraction of actin filaments. Finally, the focal adhesions at the rear of the cell are broken, perhaps by stress fiber contraction or elastic tension, so that the tail end of the cell is brought forward.

18. Ras-related G proteins are involved in the signal-transduction pathways that are activated in fibroblasts as part of the wound-healing response. Depending on the specific G protein, activation of this type of signal pathway may lead to formation of filopodia, lamellipodia, focal adhesions, and/or stress fibers. Ca^{2+} is probably involved in activation of gelsolin, cofilin, and profilin and in contraction of myosin II, and may be found in intracellular gradients important for steering in chemotactic cells.

19. Traction in cell motility is provided by focal adhesions. At focal adhesions, integrins in the cell membrane bind to the substratum. Myosin-dependent cortical contraction helps to pull the cell forward. Contraction of stress fibers in the tail may help break attachments at the rear of the cell as the cell moves forward.

ANALYZE THE DATA

a. These data show that myosin V has low ATPase activity when the free Ca^{2+} is below 1 micromolar. Thus, one would expect that myosin V would be inactive except under conditions in which the cytosolic free Ca^{2+} were elevated above this value, perhaps in localized domains of the cell or upon excitation of excitable cells.

b. These data reveal that myosin V has low ATPase activity in the absence of actin, which suggests that, in vivo, myosin V does not hydrolyze ATP at significant rates unless it can bind to and move along actin filaments. The actin-activated ATPase activity is also significantly repressed at low Ca^{2+}. Thus, two criteria may need to be met in vivo for myosin V to hydrolyze ATP. The cytosolic free Ca^{2+} concentration must become elevated above 1 micromolar, and there must be actin filaments with which the myosin V can functionally interact.

c. Unlike intact myosin V, in which the ATPase activity is inhibited in the absence of Ca^{2+}, truncated myosin V, lacking its globular tail domain, exhibits actin activated ATPase activity even in the absence of Ca^{2+}. These data suggest that the tail domain of myosin V may be required to keep the myosin in an inactive state in the absence of Ca^{2+}.

18

CELL ORGANIZATION AND MOVEMENT II: MICROTUBULES AND INTERMEDIATE FILAMENTS

REVIEW THE CONCEPTS

1. The basis of microtubule polarity is the head-to-tail assembly of αβ-tubulin heterodimers, which results in a crown of α-tubulin at the (−) end and a crown of β-tubulin at the (+) end. In nonpolarized animal cells, (−) ends are typically associated with MTOCs, and (+) ends may extend toward the cell periphery. Other arrangements occur in different types of cells, but the (−) ends are associated with a MTOC in most cases. Microtubule motors can "read" the polarity of a microtubule, and a specific motor protein will transport its cargo toward either the (+) or the (−) end of the microtubule.

2. During dynamic instability, microtubules alternate between growth and shortening. The current model to account for dynamic instability is the GTP cap model. According to this model, GTP-tubulin (subunits with GTP bound to β-tubulin) can add to the end of a growing microtubule, but at some time after assembly the GTP will be hydrolyzed to GDP, leaving GDP-tubulin, which makes up the bulk of the microtubule. Thus, GTP-tubulin is present only at the microtubule end, and as long as this situation holds, the microtubule will continue to grow since the cap stabilizes the entire microtubule. However, if GDP-tubulin becomes exposed at the end, then the stabilizing cap is lost and the microtubule will begin to shorten. The microtubule will continue to shorten until it disappears or until GTP-tubulin returns to the end and a new GTP cap is formed.

3. The best understood proteins involved in regulating microtubule assembly are the stabilizing MAPs. These proteins bind to microtubules and promote assembly, increase microtubule stability, and, in some cases, cross-link microtubules into

bundles. The other main group of MAPs functions to destabilize microtubules. This group includes proteins such as katinin, which severs microtubules, and Op18, which promotes the frequency of microtubule catastrophe.

4. Microtubule organizing centers (MTOCs), also known as centrosomes, are responsible for determining the arrangement of microtubules within a cell. The typical cell contains a single MTOC, although mitotic cells contain two (the spindle poles), and certain types of cells may contain several hundred. Microtubules are nucleated by γ-tubulin ring complexes, which are located in the pericentriolar material of an MTOC. The γ-tubulin provides a binding site for αβ-tubulin dimers, and the ring complex structure appears to provide a template for nucleating microtubule formation.

5. Microtubule/tubulin-binding drugs are used to treat a variety of diseases, including gout, certain skin and joint diseases, and cancer. Such drugs either prevent microtubule assembly (e.g., colchicine) or prevent microtubule disassembly (e.g., taxol). Although the effects are opposite, the results are the same: inhibition of cellular processes that depend on microtubules and the dynamic rearrangement of these polymers.

6. Kinesin-1 was first isolated from squid axons, which were manipulated to produce extruded cytoplasm, a cell-free system to study synaptic vesicle motility. Video microscopy was used to follow the ATP-dependent movement of synaptic vesicles along individual microtubules in the extruded cytoplasm, but when purified synaptic vesicles were added to purified microtubules, no movement was observed even in the presence of ATP. Subsequent addition of a squid cytosolic extract to the purified components restored ATP-dependent vesicle movement along microtubules, indicating that a soluble protein in the cytosol was responsible for driving vesicle movement. To identify the soluble "motor" protein, researchers took advantage of previous experiments with extruded cytoplasm that demonstrated that a nonhydrolyzable ATP analog (AMPPNP) caused vesicles to bind so tightly to microtubules that movement was stopped. To purify the motor, AMPPNP was added to a mixture of cytosolic extract and microtubules with the goal of forcing the motor to bind microtubules. The microtubules and any bound proteins were then recovered by centrifugation and treated with ATP to release proteins that bound microtubules in an ATP-dependent manner (a fundamental property of microtubule-dependent motor proteins). The predominant protein released in this approach was kinesin-1.

7. Although microtubule orientation is fixed by the MTOC (and any given motor moves only in one direction), some cargoes are able to move in both directions along a microtubule because they are able to interact with both (+)-end- and (−)-end-directed motor proteins. The direction that a given cargo moves along a microtubule appears to be controlled by swapping one motor protein for the other (it may also be possible to activate one motor and inactivate the other). Certain cargoes may move on both microtubules and actin filaments if the cargo contains binding sites for both microtubule and actin motor proteins.

8. The kinesin family motor (or head) domain contains the ATP-binding site and the microtubule-binding site of the motor, while the neck is a flexible region connecting the motor domain to the central stalk domain. The kinesin motor domain is

required to generate movement but does not appear to determine the direction a kinesin motor will move on a microtubule. Instead, the neck determines the direction of movement. These conclusions are based on experiments in which the motor domains of (+)-end- and (–)-end-directed motors were swapped with no effect on direction of motor movement, and on experiments in which mutations of the neck region caused a change from a (–)-end- to a (+)-end-directed motor. The hand-over-hand mechanism for processive movement along a microtubule requires coordinated function of two active heads. Each head must bind tightly to the microtubule in the nucleotide-free, ATP, and ADP+P_i bound states and loosely in the ADP bound state and to swing its linker region forward when it binds ATP in order for processive movement to occur.

9. The dynactin complex links cytoplasmic dynein to cargo. The weak microtubule binding site in the p150^{Glued} component of dynactin maintains association with the microtubule as the dynein motor travels down the microtubule. Inhibition of dynactin interaction with EB-1 would interfere with the spindle orientation mechanism in which dynactin interaction with the +TIP EB-1 associates dynein with a microtubule (+) end and inactivates the dynein motor activity. Growth of the microtubule to the cell cortex exposes the dynein-dynactin complex to an activator localized there that activates dynein motor activity and anchorage of the complex. The active dynein then pulls on the microtubule, which helps to orient the spindle.

10. The appendages (cilia and flagella) used for cell swimming contain a highly organized core of microtubules and associated proteins. This core, termed the axoneme, is typically made of nine outer doublet microtubules and two central pair microtubules (known as the 9 + 2 arrangement). Each outer doublet consists of a 13-protofilament microtubule and a 10-protofilament microtubule, while the central pair contains 13 protofilaments each. Cell movement depends on axoneme bending, which, in turn, depends on force generated by axonemal dyneins. These motor proteins act to slide outer doublet microtubules past each other, but this sliding motion is converted into bending because of restrictions imposed by cross-linking proteins in the axoneme, and perhaps by the action of inner arm dyneins.

11. Kinesin-2 drives microtubule (+) end-directed IFT transport to the flagellar tip. Inactivation of dynein would prevent recycling of the activated kinesin-2 back to the base of the flagellum. Inhibition of dynein activity would cause accumulation of the kinesin-2 at the flagellar tip and inhibit both anterograde as well as retrograde IFT transport.

12. The three types of microtubules that make up the spindle are kinetochore microtubules, polar microtubules, and astral microtubules. The (–) ends of all three types associate with spindle poles. Kinetochore microtubules connect chromosomes, via the kinetochore attachment site, to the spindle poles. Polar microtubules from each pole overlap and are involved in holding the poles together and regulating pole-to-pole distance. Astral microtubules radiate from each spindle pole toward the cortex of the cell, where they help position the spindle and determine the plane of cytokinesis.

13. Inhibition of kinesin-5 would be expected to have a number of effects. First, the centrosomes duplicated in S phase would not move apart during prophase to make the mitotic asters. Second, if the drug were added after the centrosomes had separated and become spindle poles, the spindle poles would not separate during anaphase B. Inhibiting kinesin-13 would affect chromosome movement of captured chromosomes in prometaphase as well as anaphase A, as the shortening of microtubules in both these cases would be compromised. Inhibiting kinesin-4—the kinesin that attaches to chromosome arms and interacts with spindle microtubules to pull them toward the center of the spindle—would affect chromosome congression at prometaphase.

14. Proteins such as kinesin-13 at the kinetochore may allow this structure to hold onto shortening microtubules. It is not clear whether this activity requires ATP hydrolysis because kinetochores have been shown to hold onto depolymerizing microtubules in vitro in the absence of ATP.

15. The separation of spindle poles during anaphase B is thought to depend on kinesin-5 motors present on microtubules in the overlap zone between the poles, which act to push the spindles apart, as well as on cytosolic dynein motors on the inner surface of the cell membrane, which act to pull astral microtubules and hence the poles apart (different organisms may utilize pushing and pulling forces to differing degrees during anaphase B). In addition, elongation of microtubules in the overlap zone appears to increase the extent of pole separation.

16. In animal cells, the spindle determines the cleavage plane. Microtubules are therefore involved with determining the plane of cytokinesis while actin filaments, as components of the contractile ring, carry out the process of cytokinesis.

17. Positive reactivity with the following monoclonal antibodies would identify the parental cell type for the types of tumors in question: a) anti-desmin, b) anti-keratin, and c) anti-GFAP protein. Each of the monoclonal antibodies should not react with cells of the other two tumor types.

18. Unlike microfilaments and microtubules, intermediate filaments do not have an intrinsic polarity. Thus, it is not surprising that there are no motor proteins that use intermediate filaments as tracks.

19. Although the mechanism for growth cone locomotion is an actin filament-dependent process similar to what drives lamellipodia protrusion in many cells, microtubules in the axon region connecting the growth cone to the cell body become stabilized by acetylation, preventing movement or collapse of the growth cone toward the cell body. The microtubules in the growth cone remain unacetylated and thus more dynamic.

ANALYZE THE DATA

1.   a. Kinesin-1 has a globular, bulbous region at one end of the molecule and this region is enlarged when antibody that binds to the motor domain is present. Accordingly, these observations suggest that the globular region contains the kinesin motor domain. The two heavy chains appear to interact in parallel, as the motor domain is observed only at one end of the molecule. In contrast,

kinesin-5 has two bulbous regions, each of which is enlarged when antibody to the motor domain is present. These data suggest that kinesin-5 has motor domains at each end of the molecule (i.e., it is a bipolar kinesin). Because it is a tetramer, rather than a dimer, it may be formed by two dimers interacting tail to tail in an antiparallel fashion.

b. The brighter end of each microtubule would represent the (–) end of the micro-tubule, whereas the less bright region would represent the (+) end of the microtubule. On kinesin-5, the microtubules are observed to glide with their (–) ends leading (bright ends lead). Accordingly, because kinesin-5 is immobi-lized on the surface and cannot move but causes the microtubule to move toward its own (–) end, relative to the microtubule kinesin-5 is moving toward the (+) end. Therefore, kinesin-5 is a (+) end microtubule motor.

c. The two microtubules are oriented with their minus ends opposite each other. Thus, because these microtubules are cross-bridged by kinesin-5, these data suggest that kinesin-5 cross-bridges microtubules of opposite polarity. When ATP is added, the ends of the two microtubules move farther apart, suggesting that kinesin-5 causes the microtubules to slide apart. Because kinesin-5 is bipolar (see text figure on page 833, part a), can cross bridge microtubules, and is a (+)-end microtubule motor (see text figure on page 833, part b), one would predict that kinesin-5 would cause two cross-bridged microtubules to slide apart with their minus ends leading.

d. Loss of Eg5 can be observed in these data to result in lack of formation of a bipolar mitotic spindle. Therefore, Eg5 appears to be required to form a nor-mal spindle. Because kinesin-5 can induce sliding apart of microtubules of opposite polarity (see text figure part c), Eg5 might be required to help the two centrosomes move apart to form the poles of a bipolar spindle. If the two poles (centrosomes) do not move apart in preparation for mitosis, then a bipo-lar spindle cannot form between them.

2   a. Both the full-length (GST-VIM-FL) vimentin and the vimentin head domain (GST-VIM-Head) GST fusion proteins pull down AKT1, but the vimentin coiled-coil domain (GST-VIM-CC) and the tail (GST-VIM-TAIL) regions do not (left). The full length AKT1 and the AKT1 tail domain pull down vimentin, but the PAH domain (GST-AKT1-PH) and the catalytic (GST-AKT1-CAT) domains do not. This demonstrates that the vimentin head domain interacts directly with the AKT1 tail domain.

b. Western blot analysis with the PAS antibody that reacts with sites phos-phorylated by AKT1 demonstrates that AKT1 phosphorylates the wild type and S325A mutated vimentins but not the S39A vimentin mutation. This identifies vimentin Serine 39 as the site of AKT1 phosphorylation.

c. Expression of active AKT1 (AKT1DD) has no affect on cell migration, but over-expression of vimentin alone (VIM) significantly increases cell migration and that migration is further enhanced by expressing both vimentin and active AKT1 (VIM+AKT1DD). This enhancement is blocked by substituting the wild-type vimentin with the mutation that cannot be phosphorylated by AKT1 (VIMS39A +AKT1DD). Expression of the phosphomimetic form of the vimen-tin (VIMS39D), even in the absence of AKT1, mimics the effect of AKT1 phos-phorylation of wild-type vimentin. These results indicate that phosphorylation of vimentin S39 or mimicking that phosphorylation maximizes cell migration.

19

THE EUKARYOTIC CELL CYCLE

REVIEW THE CONCEPTS

1. The unidirectional and irreversible passage through the cell cycle is brought about by the degradation of critical protein molecules at specific points in the cycle. Examples are the proteolysis of securin at the beginning of anaphase, proteolysis of cyclin B in late anaphase, and proteolysis of the S-phase CDK inhibitor at the start of S phase. The proteins are degraded by a proteasome, a multiprotein complex. Proteins are marked for proteolysis by the proteasome by the addition of multiple molecules of ubiquitin to one or more lysine residues in the target protein. Securin and cyclin B are both poly-ubiquitinylated by the APC/C complex. The S-phase CDK inhibitor is polyubiqui-tinylated by SCF.

2. Experimental strategies employed to study cell cycle progression include use of live whole multicellular organisms such as *D. melanogaster* and *Xenopus* embryos, living unicellular subsystems such as *Xenopus* oocytes and yeast cells (*S. pombe* and *S. cerevisiae*), and mammalian cell culture systems. Using biochemical strategic approaches the environment within which these systems are maintained can be manipulated by the researcher to simulate both growth and non-growth situations. Radioactive nucleic or amino acids can be added to the media to study either gene or protein expression. Extracts from each of these systems can also be obtained at the desired time and then probed with either antibodies or nucleic acid probes.

    Genetic experimental approaches would involve manipulating the genome of the system under study and then evaluating the phenotype of the resulting mutant(s), perhaps using the biochemical approaches outlined above. In the case of lethal mutations, cDNA libraries can be employed to identify and investigate potential genes that may "rescue" such lethal mutants.

3. Hunt and colleagues used a variety of marine invertebrate cell culture systems in their discovery of cyclin proteins. The experimental strategy employed was primarily a biochemical one, in which radiolabeled amino acids were added to the cell culture media under controlled conditions. Isolation of cell extracts at various time points under various conditions enabled the researchers to differentiate between proteins expressed in both the pre- and post-fertilization states of development. Through correlating the presence or absence of several proteins with the cell cycle stage identified through light microscopy in several cell culture systems, Hunt and colleagues were able identify the cyclin protein family.

4. Murray and Kirschner performed a classic set of experiments in frog egg extracts to reveal the essential role of cyclin B synthesis and degradation in cell-cycle progression. In one experiment, extracts were treated with RNase to destroy all endogenous mRNAs. These extracts arrested in interphase, suggesting that an essential protein needed to be translated to drive the cell cycle into mitosis. When RNase-treated extracts were supplemented with a single exogenous mRNA encoding wild-type cyclin B, the extract progressed into mitosis, indicating that cyclin B was the essential protein that had to be synthesized to drive entry into mitosis. When RNase-treated extracts were supplemented with mRNA encoding a nondegradable form of cyclin B, the extract entered mitosis and arrested there with high MPF activity instead of eventually destroying MPF and exiting mitosis. This experiment revealed that degradation of cyclin B was necessary for mitotic exit.

5. Both strains of yeast are eukaryotic and can (although do not always) exist in the haploid state, meaning that they carry only one copy of their genome. Both can be easily cultured and utilized to develop mutant strains that contain alterations in single genes. In addition, the genomes of both strains have been sequenced and characterized. However, these yeast strains are very different from each other from both genotypic and phenotypic perspectives and individual genes can be studied independently in both yeast systems.

6. The inhibitory tyrosine-phosphate in the T-loop of mitotic CDK is the substrate of Cdc25. MPF activates Cdc25 (by phosphorylating it) and Cdc25, in turn, activates MPF (by dephosphorylating it). Therefore, when a small amount of active MPF is injected into an immature egg, it will phosphorylate and activate the endogenous Cdc25, which can then dephosphorylate and activate MPF. This explains the autocatalytic nature of MPF.

7. a. CDKs are active as kinases only when bound to a cyclin. Cyclin binding exposes the active site of the CDK and also helps to form the substrate-binding pocket.

   b. CAK is a kinase that phosphorylates cyclin-CDKs on a threonine residue in the T loop. This phosphorylation induces a conformational change that increases affinity of CDKs for their substrates, thereby greatly enhancing the catalytic activity of the CDK.

    c. Wee1 is a kinase that phosphorylates CDKs on tyrosine 15 in the ATP-binding region. This phosphorylation interferes with ATP binding and thereby inhibits the catalytic activity of the CDK.

    d. p21 is a stoichiometric inhibitor that binds and inhibits the activity of cyclin-CDKs, usually in response to damaged DNA.

8. Mitogen-dependent progression through the first gap phase (G_1) and initiation of DNA synthesis (S phase) during the eukaryotic cell division cycle are cooperatively regulated by several classes of cyclin-dependent kinases (CDKs) whose activities are in turn constrained by CDK inhibitors (CKIs). CKIs inhibit the function of CDKs. CKIs that govern these events have been assigned to one of two families based on their structures and CDK targets. The first class includes the INK4 proteins (inhibitors of CDK4), so named for their ability to specifically inhibit the catalytic subunits of CDK4 and CDK6. The second class of CKIs inhibit G_1/S-CDKs and S-Phase CDKs, and must be degraded before DNA replication can begin. The activity of CKIs represents an additional layer of cell-cycle regulation over and above that mediated by temporal cyclin synthesis and degradation.

9. START represents a point in the cell cycle beyond which cells are irrevocably committed to completing DNA replication and mitosis even if growth factors, or mitogens, are removed. To enter the cell cycle, quiescent cells in G_0 require growth factors, which bind to cell-surface receptors and trigger a signaling cascade that leads to the transcription of early-response genes and then delayed-response genes. Among the delayed response genes is the cyclin D gene, which partners with CDKs 4 and 6, and this mid-G_1 cyclin-CDK complex phosphorylates the Rb protein. When Rb is phosphorylated by the mid-G_1 cyclin-CDK, it can no longer bind the transcription factor E2F. When E2F is released from Rb, then it induces transcription of the genes that promote entry into S phase.

    a. High levels of cyclin D bypass the requirement for growth factors, which normally induce synthesis of cyclin D.

    b. If Rb is not functional, then growth-factor induced synthesis of cyclin D is not required to promote phosphorylation and inactivation of Rb by CDK4/6. E2F will be constitutively active.

    c. p16 (INK4A) is an inhibitor of G_1/S-CDK complexes. If p16 function is lost, this promotes activation of G_1/S-CDK complexes and entry into S phase.

    d. E2F proteins are transcription factors that stimulate expression of genes whose products are required for entry and progression through S phase. E2F activity also stimulates its own expression, and E2F hyperactivity in this regard would further promote entry into S phase.

10. Unphosphorylated Rb protein binds to E2F proteins, repressing transcription of genes for various proteins required for the S phase. When Rb is phosphorylated by the mid-G_1 cyclin-CDK, E2F is liberated. E2F activates transcription of genes required for entry into S phase. Two of these genes code for cyclin E and CDK2, the late G_1 cyclin-CDK, which further phosphorylates Rb in a positive feedback loop. In this way, Rb remains phosphorylated throughout the S, G_2, and early M phases.

11. G_1 cyclin-CDKs phosphorylate and target Sic1 for degradation, which releases active S-phase cyclin-CDK complexes. They also inactivate the APC/C by phosphorylation of Cdh1, which allows B-type cyclins to accumulate. G_1 cyclin-CDKs promote the synthesis of mitotic cyclins by activating their transcription factor, MBF.

12. In *S. cerevisiae*, S-phase cyclin-CDKs become active at the beginning of S, when the CDK inhibitor Sic1 is degraded. These S-phase cyclin-CDKs, as well as other mitotic cyclin-CDKs synthesized later in S and in G_2, remain active until late anaphase. Prereplication complexes can assemble on origins of replication only during G_1, when mitotic cyclin-CDK activity is low. Origins are activated by recruitment of other initiation factors to the pre-RC to form the pre-initiation complex (pre-IC) and this event requires CDKs and Dbf4Dependent Kinase, Dbf4-Cdc7 (DDK). These events lead to activation of the Mcm2-7 helicase and recruitment of DNA synthesis machinery. The S-phase kinase, DDK, phosphorylates the Mcm2-7 helicase, and is thought to be required either for helicase activation or for recruitment of pre-IC factors (or both). Once an origin has "fired" (i.e., replication has been initiated), the persistence of mitotic cyclin-CDK activity during S, G_2, and M prevents reassembly of prereplication complexes on that origin until the cell has completed the segregation of chromosomes in late anaphase and mitotic cyclins are degraded. Therefore, each origin initiates replication once and only once per cell cycle because of the oscillating activity of mitotic cyclin-CDKs.

13. The wee phenotype in *S. pombe* displays smaller than usual cells. Premature entry into mitosis, before the cell has grown to the size that normally signals cell division, is the cause of this phenotype. Wee cells result from the excess activity of Cdc2, the cyclin-dependent kinase of *S. pombe* MPF. The wee phenotype can result from a mutation in the *wee1+* gene, which encodes the Wee1 protein kinase responsible for catalyzing the addition of phosphate to tyrosine 15 of Cdc2, which inhibits Cdc2 function and prevents premature entry into mitosis. The wee phenotype also results when a mutation renders Cdc2 insensitive to Wee1 or a mutation in which Cdc25, the phosphatase that opposes Wee1, is overexpressed. Discovery of the wee phenotype and the characterization of the *wee1+* gene revealed the intimate link between cell size and cell-cycle progression as well as the important role that tyrosine phosphorylation plays in regulating the activity of CDKs.

14. Monitoring of sister kinetochore attachment to the mitotic spindle is accomplished through a mitotic spindle assembly checkpoint (SAC). The spindle assembly checkpoint (SAC) is an active signal produced by improperly attached kinetochores, and is highly conserved in all eukaryotes. The SAC stops the cell cycle by negatively regulating Cdc20, thereby preventing the activation of the polyubiquitylation activities of anaphase promoting complex (APC). The proteins responsible for the SAC signal compose the mitotic checkpoint complex (MCC), which includes SAC proteins and Cdc20.

15. To initiate sister chromatid segregation at anaphase, the APC/C polyubiquitinylates securin, targeting it for degradation by the proteasome. Degradation of

securin releases the enzyme separase, which cleaves kleisin, a component of the cohesin complexes that hold sister chromatids together.

16. a. Ime2 replaces the G_1-CDK function of phosphorylating Sic1, allowing the cell to enter S phase. The cell cannot use the normal G_1 cyclin-CDKs for this because diploid yeast cells are induced to enter meiosis by starvation. In the absence of nutrients, the mid- and late-G_1 cyclins are not expressed and consequently cannot function. In contrast, Ime2 is induced by starvation. Since Ime2 is expressed during meiosis I but not meiosis II, DNA replication is prevented during meiosis II, allowing for reduction to $1n$ chromosome content in the resulting gametes.

  b. Rec8, a homolog of the mitotic cohesin subunit kleisin, maintains centromeric cohesion of sister chromatids during meiosis I. Centromeric Rec8 is protected from degradation by separase during meiosis I so that sister chromatids remain attached. Rec8 is degraded during meiosis II when sister chromatids must separate.

  c. Monopolin is required for the formation of specialized kinetochores during meiosis I that co-orient sister chromatids of synapsed homologous chromosomes so that they bind to microtubules emanating from the same spindle pole.

17. A cell-cycle checkpoint is a place in the cell cycle where a cell's progress through the cycle is monitored, and, if the current process has not been completed properly, further progression through the cell cycle is inhibited; the cell cycle is arrested at this checkpoint until the process in question is completed successfully. Checkpoint pathways exist at G_1 and S phases to assess DNA damage, at G_2 to assess DNA damage and to determine whether DNA replication is complete, and at M phase to identify any problems with assembly of the mitotic spindle or chromosome segregation. Because these checkpoint pathways identify problems with the genome (unreplicated, damaged, or improperly segregated DNA) and arrest the cell cycle so that these problems can be fixed, checkpoint pathways can prevent the propagation of mutations into the next cell generation and thereby preserve the fidelity of the genome.

18. In a normal cell, p53 is rapidly degraded as a result of polyubiquitination by Mdm2, a ubiquitin-protein ligase. DNA damage activates ATM and ATR, which phosphorylate p53, blocking the interaction with Mdm2 and therefore interfering with p53 degradation. As a consequence, the p53 concentration increases in cells with DNA damage. p53 acts as a transcription factor for several genes involved in protecting cells from DNA damage. One of these genes codes for p21^{CIP}, a cyclin/CDK inhibitor. p21^{CIP} inhibits all mammalian cyclin-CDK complexes and, as a result, the cell cycle cannot be completed until the DNA damage is repaired and phosphorylation of p53 by ATM and ATR ceases.

19. Besides p53, ATM also phosphorylates Chk1 and Chk2 kinases (as well as several additional substrates). Chk1 and Chk2 phosphorylate the phosphatases Cdc25A and Cdc25C, targeting Cdc25A for degradation and inactivating Cdc25C. In the absence of Cdc25 phosphatases, CDKs are maintained with inhibitory phosphorylations, thereby arresting the cell cycle.

ANALYZE THE DATA

1. a. Cyclin B is degraded more quickly when Xnf7 is depleted from the extracts. Thus, these studies suggest that Xnf7 functions in some way to delay cyclin degradation and perhaps the onset of anaphase. Because Xnf7 binds to APC/C, it is possible that Xnf7 normally inhibits APC/C. If so, depletion of Xnf7 would allow APC/C to be activated and target cyclin B for destruction sooner, as observed here.

   b. In untreated extracts, cyclin B is polyubiquitinylated 10 minutes after release from metaphase arrest, whereas addition of exogenous Xnf7 delays the onset of cyclin polyubiqutination to 16 minutes. These data reinforce those in (a), suggesting that the presence or absence of Xnf7 affects the timing of cyclin B ubiquitinylation and subsequent destruction. These data suggest that Xnf7 affects the activity APC/C, the ligase responsible for ubiquitinylating cyclin B.

   c. Cells proceeding normally through mitosis (see text figure on page 921, top panel) degrade cyclin at 40 minutes after addition of Ca^{2+}. When the spindle checkpoint is activated, as in the presence of nocodazole (middle panel), cells are checked and cyclin B is not degraded as it is in control cells (top panel). However, if the extracts are depleted of Xnf7 (bottom panel), then cyclin B is degraded even though the cells are in nocodazole and should be checked at a stage prior to cyclin degradation. These data suggest that Xnf7 is required for maintenance of the spindle checkpoint.

2. a. Northern blots could be utilized as a first step to evaluate whether cyclin B expression is regulated at the transcriptional level. Western blots could be utilized in a similar fashion to evaluate the role of cyclin B translation. In both cases, cellular extracts could be isolated at various points in the cell cycle and subsequently analyzed using autoradiography in the case of a Northern blot or labelled antibodies in the case of a Western blot.

   b. Yes—cyclin B could be regulated at the post-translational level through phosphorylation at specific amino acid sites. In addition, cyclin B activity could be regulated through other post-translational modifications including glycosylation, carboxylation, methylation or acylation. Moreover, the activity of cyclin B could be regulated through localization or compartmentalization within the cell. Signal-transduction mechanisms based on the cell's ability to react to its external environment could play a role in these events.

   c. Cyclin B is a mitotic cyclin. Its expression could be mediated by events in the cell's external environment that promote cellular proliferation, for example, in the presence of growth hormones. In these cases, the cell would have to recognize that DNA replication has already taken place. One potential manner by which this could be regulated is through expression of specific markers on the cell surface that signal that S phase is complete and that M phase is about to commence. Signal-transduction mechanisms could be regulated by external peptides or other moieties that interact either directly or indirectly with these specific G_2 phase cell-surface markers. Such signal-transduction mechanisms could promote either the synthesis or activity of cyclin B.

20

INTEGRATING CELLS
INTO TISSUES

REVIEW THE CONCEPTS

1. The diversity of adhesive molecules has arisen from 1) duplication of a common ancestor gene followed by divergent evolution producing multiple genes encoding related isoforms; and 2) alternative splicing of a single gene to yield many mRNAs, each encoding a distinct isoform. Integrins are heterodimers of α and β subunits. Combinatorial diversity of 18 α and 8 β subunits yields at least 24 functional integrin heterodimers.

2. Homophilic interactions are those between like cell types (e.g., epithelial cells with epithelial cells). One approach to demonstrating homophilic cell interactions experimentally is to use L cell lines transfected with E-cadherin. L cells adhere poorly to each other and express no cadherins. When transfected with E-cadherin, L cells adhere tightly through homotypic E-cadherin interactions among cells. Ca^{2+} is required for cadherin interaction. Removing extracellular Ca^{2+} disrupts interaction between the E-cadherin expressing cells.

3. Actin and myosin filaments form a circumferential belt in a complex with adherens junctions. This belt functions as a tension cable that can internally brace the cell and control its shape.

4. Tight junctions define apical and basolateral plasma membrane domains in polarized epithelial cells and control the flow of solutes between cells in an epithelial sheet. Several things can happen to cells when tight junctions do not function. In hereditary hypomagnesemia, defects in tight junctions prevent the normal flow of magnesium

through tight junctions in the kidney. Low blood-magnesium levels result, which can lead to convulsions. Altering tight junctions in hair cells of the cochlea of the inner ear can result in deafness.

5. Gap junctions can be electrical synapses through which ions pass rapidly from cell to cell in either direction, unlike chemical synapses. The electrical synapses between cardiac muscle cells helps coordinate beating. Although gap junction communication can be regulated by pH, Ca^{2+} concentration, phosphorylation, and voltage-gating, myometrial connexin Cx43 expression is upregulated to increase the number and size of gap junctions for parturition and decreases rapidly postpartum.

6. Collagen is a major component of the extracellular matrix in animal cells. It is a protein that has a trimeric structure with rodlike and globular domains that form a two-dimensional network. Collagen is synthesized in its precursor form by ribosomes attached to the endoplasmic reticulum. These pro-α chains undergo a series of covalent modifications and are folded into a triple-helical procollagen molecule. The folded procollagen is transported through the Golgi and the chains are secreted to the outside of the cell. Once outside the cell, peptidases cleave the N- and C-terminal propeptides. The triple helices are then able to form larger structures called collagen fibrils. Mutations in collagen IV or autoimmune antibodies that disrupt collagen IV are associated with progressive renal failure, hearing loss, lung hemorrhage, and sight abnormalities.

7. Structural studies have shown that integrin exists in both a non-active, low-affinity, or "bent," form and an active, high-affinity, or "straight," form. In outside-in signaling, molecules of the ECM can bind to the extracellular portion of inactive integrin and induce conformational changes that lead to the straightening of the intracellular, cytoplasmic tails of integrin. The straightening of the cytoplasmic tails can stimulate intracellular components such as the cytoskeleton and parts of signaling pathways. This structure also facilitates inside-out signaling. For example, when the metabolic state of the cell is altered, adapter proteins inside the cell can interact with the cytoplasmic tails of integrin and cause either straightening or association. This would result in straightening or bending of the integrin extracellular domains and either promote or inhibit interaction of the integrin with the ECM.

8. Proteoglycans are highly viscous glycoproteins that cushion cells and bind to a wide variety of extracellular molecules. Collagen is fibrous and provides structural integrity, mechanical strength, and resilience. Soluble extracellular matrix proteins such as laminin and fibronectin bind and cross-link cell-surface receptors and other ECM components.

9. Syndecans in the hypothalmic region of the brain participate in the binding of antisatiety peptides to cell-surface receptors. In the "fed" state, the extracellular domain of syndecans is released from the surface by proteolysis. When this happens, the activity of antisatiety peptides is suppressed along with feeding behavior.

10. The RGD sequence on fibronectin mediates binding to integrin proteins. If RGD-containing peptides were added to a layer of fibroblasts grown on a fibronectin substrate in tissue culture, the RGD peptides would compete with fibronectin for binding to the integrins present in the fibroblast extracellular matrix. As a result, the fibroblasts would likely lose adherence to the fibronectin substrate.

11. Three types of proteases can degrade ECM components as well as non-ECM components such as surface adhesion receptors: MMPs (matrix metalloproteases), ADAMs (a disintegrin and metalloproteinases), and ADAMTS (ADAM with thrombospondin motifs). These proteases can be integral membrane proteins or secreted proteins, some of which bind tightly to membrane receptors. ECM degrading proteases are associated with a variety of diseases, including metastatic cancer.

12. Fibronectin contains RGD- and fibrin-binding domains. Binding of fibronectin to a fibrin clot recruits platelets through interaction of the fibronectin RGD domain with a platelet integrin.

13. The dystrophin gene, which is defective in Duchenne muscular dystrophy, is an adapter protein that binds to cytoskeletal components such as actin as well as to the cell-adhesion molecule dystroglycan. Normally, dystrophin and dystroglycan function in an important part of the signaling relay linking the extracellular matrix on the outside of the muscle cell to the cytoskeleton and signaling components inside the muscle cell. When any of these components is defective, the muscle cells do not develop or function properly and muscular dystrophy results.

14. The process is called extravasation. Inflammatory signals including chemokines are released in the area of infection. These signals activate the endothelial cells lining blood vessels in the area. P-selectin exposed on the surface of activated endothelial cells mediates weak adhesion of passing leukocytes. Weakly bound leukocytes roll along the surface of the endothelium. At the same time, chemokines and other signaling molecules including the platelet-activating factor (PAF) also activate $\beta2$-containing integrins on the cell surface of the leukocytes. Upon activation, the integrins change conformation in to their high-affinity form. Activated integrins bind to IgCAMs on the surface of endothelial cells. Tightly bound leukocytes stop rolling, spread out on the surface of the endothelium, and eventually crawl between adjacent endothelial cells into the underlying tissue.

15. Small molecule hormones, called auxins, induce the weakening of the cell wall. This permits expansion of the intracellular vacuole by uptake of water, leading to cell elongation.

16. Both plasmodesmata and gap junctions are channels that directly connect the cytosol of one cell to that of an adjacent cell. However, in plasmodesmata, the plasma membranes of the adjacent cells are merged to form a continuous channel, the annulus. Membranes of the cells at a gap junction are not continuous.

Plasmodesmata may also contain an extension of the endoplasmic reticulum, the desmotube, that passes through the annulus. Animal cells do not contain desmotubes.

ANALYZE THE DATA

a. Cells transfected with wild-type E-cadherin aggregate more than untransfected cells because the increase in E-cadherin allows for more cell-cell interactions via the ECM.

b. Mutant A behaves almost identically to the wild-type E-cadherin in the aggregation assay; therefore, this mutation does not change the function of E-cadherin as far as hemophilic interactions are concerned. Expression of mutant B, however, does not result in increased aggregation, so this particular mutation does alter the adhesive qualities of E-cadherin.

c. The monoclonal antibody specific for E-cadherin blocks the resulting aggregation because it binds to E-cadherin and blocks homophilic interactions. In contrast, the nonspecific antibody does not specifically interact with E-cadherin, and does not block homophilic interactions.

d. Since cadherins require calcium for function, lowering the calcium in the media during the assay would lower the aggregation ability.

21

STEM CELLS, CELL ASYMMETRY, AND CELL DEATH

REVIEW THE CONCEPTS

1. By definition, a stem cell divides to give rise to a copy of itself and to a differentiated cell or a cell capable of differentiating into multiple cell types, such as a mutipotent progenitor cell. Totipotent stem cells can give rise to every tissue in an organism. Pluripotent stem cells give rise to multiple, but not necessarily all, cell types. Progenitor cells give rise to more than one cell type but, unlike stem cells, do not self-renew.

2. In plants, stem cells are located in meristems, such as shoot apical meristems (SAMs) and floral meristems. In adult animal cells, stem-cell populations are thought to exist in low numbers in many organs including skin, intestine, and bone marrow. SAMs in plants are embryo-like in their concentration of totipotent stem cells. However, stem cells are difficult to purify from adult animals, and the only totipotent stem cells found in animals are in very early stage embryos.

3. Because Dolly was derived from an egg containing a nucleus from an adult, differentiated cell, we know that differentiated nuclei (at least adult mammary cells) have the potential to dedifferentiate and become totipotent. Since Dolly was derived from a differentiated nucleus placed into an egg and not from an intact, differentiated cell, we can conclude nothing about the ability of a differentiated cell to become totipotent. Other than the nucleus, the organelles, including mitochondria, which also contain genetic material, were derived from a germ cell. Differentiation of cells is maintained by cytoskeletal structures, organelles that confer cell properties, particular modifications of key regulatory proteins, and accessibility of regulatory genes in the chromatin.

The Dolly experiment best indicates that the chromatin of differentiated nuclei can be remodeled from a differentiated to a totipotent state.

4. (a) pluripotent cells, (b) pluripotent cells, (c) totipotent cells, (d) multipotent cells.

5. True. Somatic cell nuclear transfer (SCNT) where the nucleus of an adult somatic cell is introduced into an enucleated egg to produce the equivalent of a zygote can, in rare cases, develop into a fully functional organism even though the source of the DNA is a differentiated somatic cell from an adult. Furthermore, induced pluripotent stem cells (from adult somatic cells) can be experimentally introduced into a blastocyst and form all of the tissues of a mouse, including germ cells.

6. Stem cell identity is maintained by Wnt signaling via β-catenin activation. Over-production of active β-catenin in intestinal cells leads to excess proliferation of the intestinal epithelium, whereas blocking the function of β-catenin depletes stem cells in the intestine.

7. **a** cells secrete only **a**-type mating pheromone and express only α-type pheromone receptor. Therefore, each haploid cell is able to attract and respond only to cells of the opposite mating type.

8. Because *C. elegans* consists of a small, invariant number of cells, it has been possible to generate a fate map of every cell from the fertilized egg to adulthood. *C. elegans* is also very amenable to genetic manipulation. Therefore, it is possible to alter the expression of specific genes and then determine the effect of this manipulation on cell division, cell differentiation, and cell death. Because many differentiation pathways are highly conserved between *C. elegans* and mammals (e.g., the apoptotic pathway), much of the information derived from studies in *C. elegans* can be applied by analogy to mammalian systems and homologous genes can be discovered.

9. In *S. cerevisiae*, the myosin motor protein, Myo4p, localizes Ash1 mRNA to the bud that will form the daughter cell. In *Drosophila* neuroblasts, microtubules are required for assembly of the Baz/Par6/PKC3 protein complex at the apical end.

10. A complex comprising Par3, Par6, and aPKC is localized anteriorly in the developing embryo and blocks expression of members of another complex (containing Par1 and Par2) from that region of the embryo. Conversely, the Par1/Par2 complex is expressed posteriorly within the embryo and prevents members of the Par3/Par6/aPKC complex from being expressed.

11. In mutant mice in which either neurotrophins or their receptors are knocked out, specific classes of neurons die by apoptosis. These results indicate that apoptosis occurs by default unless a specific extracellular signal is transduced to block the apoptotic program.

12. In apoptosis, cells shrink and condense before being phagocytosed by macrophages. Apoptosis does not expose neighboring cells to potentially toxic

substances. In necrosis, cells swell and burst, emptying their contents into the surroundings. Necrosis triggers a potentially damaging inflammatory response in the tissue.

13. Killer proteins initiate the events surrounding apoptosis. Destruction proteins digest cell components such as DNA. Engulfment proteins aid phagocytosis of dying cells.

14. (a) no cell death, (b) no cell death, (c) cell death.

15. Although external signals such as TNF and Fas ligand induce apoptosis, the responding cell must still transduce the death signal through an intracellular pathway and induce its own death by activation of the caspase enzymes. The morphologic events of this death are indistinguishable from apoptosis triggered by an intrinsic pathway.

16 a. The cell should undergo apoptosis even in the presence of trophic factors.

b. The cell should not undergo apoptosis even in the absence of trophic factors.

c. The cell should not undergo apoptosis even in the absence of trophic factors. The mutations named in the question in both (b) and (c) could be found in cancer cells because either would block apoptosis even in the absence of trophic factors.

17. IAPs have zinc-binding domains that can bind directly to caspases and inhibit their protease activity, thus preventing apoptosis. SMAC/DIABLOs, a family of mitochondrial proteins, can bind to the zinc-binding domains in IAPs and prevent them from binding to caspases.

ANALYZE THE DATA

a. The lower panel in the text figure (which can be found at the *Molecular Cell Biology* website: www.whfreeman.com/lodish7e) shows a cell that appears to have differentially segregated most of its BrdU-labeled DNA strands to only one of its two daughter cells. Because the older DNA strands would contain BrdU and the newer strands, synthesized during the 18-hour period in the absence of BrdU (chase), would not contain BrdU, this observation suggests that older (labeled) strands have been co-segregated to the lower daughter while the upper daughter received few or no detectable BrdU-labeled strands. In contrast, both daughter cells in the upper panel contain BrdU-labeled strands. The asymmetric partitioning of 40 labeled chromosomes, observed here to occur with a frequency of about 1.5 in 10^2 after only two divisions in the absence of BrdU, would be expected to occur by chance at a frequency of about $(1/2)^{39}$ or about 1.8 in 10^{12} (i.e., the loss of label from one of the two daughter cells after two divisions is occurring about 10 billion times more frequently than expected).

b. The cell that acquires most of the Numb also acquires the older (BrdU-labeled) DNA strands and thus may become the new stem cell. To determine if Numb is required for co-segregation of older DNA strands, one could deplete satellite cells of Numb by using a siRNA directed against Numb mRNA and then examining the progeny to determine if any cells exhibit co-segregation of older DNA strands.

c. The established cell line has none of the properties of stem cells and apparently has lost the ability to mark or detect the older DNA strands at each division and thereby co-segregate them preferentially to one of the two daughter cells. Only symmetric cell divisions, with random segregation of the DNA strands, would be expected.

22

NERVE CELLS

REVIEW THE CONCEPTS

1. Four types of glial cells interact with neurons. Oligodendrocytes and Schwann cells form the myelin sheath around neural axons in the central nervous system and peripheral nervous system, respectively. Interactions between glia and neurons control the placement and spacing of myelin sheaths, and the assembly of nerve transmission machinery at the nodes of Ranvier. Another type of glial cell, astrocyctes, is important in producing synapses, forming contacts with synapses, and producing extracellular matrix proteins. Microglia produce survival factors and carry out immune functions.

2. The negative resting potential in animal cells is generated by the action of the Na^+/K^+ ATPase. This pump uses the energy of ATP hydrolysis to move Na^+ ions outside the cell and K^+ ions into the cell. Three Na^+ ions are pumped out for every two K^+ ions pumped into the cell. Thus, operation of the Na^+/K^+ pump generates a high-K^+ and low-Na^+ concentration inside the cell relative to the K^+ and Na^+ concentrations in the extracellular medium. This concentration gradient drives the movement of K^+ ions across the plasma membrane to the outside of the cell through nongated potassium channels, generating the resting membrane potential.

3. An action potential has three phases: depolarization, in which the local negative membrane potential goes to a positive membrane potential; repolarization, in which the membrane potential goes from positive to negative; and hyperpolarization, in which the resting negative membrane potential is exceeded. Depolarization corresponds to the opening of voltage-gated Na^+ channels and a resulting influx of Na^+.

Repolarization corresponds to the opening of voltage-gated K$^+$ channels. Hyperpolarization corresponds to a period of closure and inactivation of voltage-gated Na$^+$ channels. The Na$^+$ channels involved in the propagation of an action potential are referred to as voltage-gated channels because they open only in response to a threshold potential.

4. The voltage-sensing domains of voltage-gated potassium channel proteins are "arms" or "paddles" that protrude into the surrounding membrane from a central transmembrane pore formed by helices S5 and S6 from each of the four identical subunits in potassium ion channels. The voltage-sensing domains of the channel proteins involve helices S1–S4. S4 has a positively charged lysine or arginine every third or fourth residue. Changes in voltage cause the helices in the voltage domain to move across the membrane. Movement of voltage domain exerts a torque on a linker helix that connects S4 to S5. This causes the S5 helix to move in such a way that the central pore is either squeezed closed or opened. The N-terminal part of each subunit of the shaker potassium channel protein forms a "ball" that extends into the cytosol and serves as the channel inactivating segment. This domain tucks into a hydrophobic pocket as long as the channel remains open but moves to re-close the channel within milliseconds after the channel opens. All voltage-gated channels are thought to have evolved from a monomeric ancestral channel protein that contained six transmembrane α helices (S1–S6). Furthermore, all voltage-gated ion channels are thought to function in a similar manner. Thus, even though some of the structural details are different, many of the biochemical insights revealed by protein crystal structures of 75 potassium channels apply to other ion channels as well.

5. Voltage-gated Na$^+$ and K$^+$ channels are clustered at the nodes of Ranvier. When the membrane potential increases at one node, the Na$^+$ channels at the next node "feel" the increased positive voltage. They open, causing sodium ions to flood into the cell at that point. The membrane potential increases and the Na$^+$ channels at the next node "feel" the increased voltage. This continues until the action potential reaches the axon terminal.

6. As membrane potential approaches E$_{Na}$, Na$^+$ channels become inactivated, preventing additional influx of Na$^+$. K$^+$ channels open, facilitating an efflux of K$^+$ from the cell.

7. Once the threshold potential to start an action potential is reached, a full firing occurs. The signal information is therefore carried primarily not by the intensity of the action potentials, but by the timing and frequency of them.

8. Hyperpolarization that results from opening voltage-gated K$^+$ channels delays opening of Na$^+$ channels upstream. Sodium channels are also inactive for a few milliseconds after an action potential has passed. This refractory period prevents a nerve impulse from traveling backward.

9. During the refractory period, Na$^+$ channels are inactivated and are therefore unable to transport Na$^+$ ions across the membrane to depolarize the cell.

10. Myelination is the development of a myelin sheath about a nerve axon. The myelin sheath is an outgrowth of neighboring glial (Schwann) cell plasma membrane that repeatedly wraps itself around the neural extension until all the cytosol between the layers of membrane is forced out. The remaining membrane is the compact myelin sheath. The myelin sheath serves as an insulator around the axon and hence speeds the rate of action potential propagation tenfold to a hundredfold. The myelin sheath surrounding an axon is formed from many glial cells. Between each region of myelination is a gap, the node of Ranvier. The voltage-gated Na^+ channels that generate the action potential are all located in the nodes. The action potential spreads passively through the axonal cytosol to the next node. This produces a situation in which the action potential in effect jumps from node to node. If the nodes are located too far apart, for example, a tenfold increase in spacing, then the passive spread of the action potential may become too slow to jump from node to node.

11. Under normal circumstances, leftover neurotransmitters released into the synaptic cleft are quickly removed from this location via reuptake or targeted degradation. Cocaine binds to and inhibits the re-uptake transporters for norepinephrine, serotonin, and dopamine. As a consequence, a higher than normal concentration of these neurotransmitters (especially dopamine) remains in the synaptic cleft, prolonging the stimulation of postsynaptic neurons. This extended stimulation leads to adaptation and the down-regulation of dopamine receptors and thus altered regulation of dopaminergic signaling, which is why habitual users tend to need to use more and more in order to achieve the same high.

12. Both the accumulation of neurotransmitters in synaptic vesicles and the accumulation of sucrose in the lumen of the plant vacuole are mediated by H^+-linked antiporters. In both cases, the proton moves down an electrochemical gradient to power the inward movement of the small organic molecule against a concentration gradient. Acetylcholine (as a released neurotransmitter) is degraded by acetylcholine esterase after its release into the synaptic cleft. Decreased acetylcholine esterase activity at the nerve-muscle synapse has the effect of prolonging signaling and hence prolonging muscle contraction.

13. The resting potential of the muscle plasma membrane (which is now permeable to Na^+ and K^+) is near E_K, so the opening of acetylcholine receptor channels in response to motor neuron stimulation causes little increase in the efflux of K^+ ions, but Na^+ ions rush into the cell (down their gradient) and depolarize the membrane. The shift in membrane potential triggers opening of voltage-gated Na^+ channels, leading to further depolarization and the generation and conduction of an action potential in the muscle cell surface membrane stimulating the release of Ca^{2+} from sarcoplasmic reticulum stores.

14. Such rapid fusion of synaptic vesicles with plasma membrane in response to Ca^{2+} influx strongly indicates that the fusion machinery is assembled in a resting state and can rapidly undergo a conformational change. Synaptic vesicles loaded with neurotransmitter are localized near the presynaptic plasma membrane. Some synaptic vesicles are "docked" at the plasma membrane; others are in reserve in the active zone near the plasma membrane. In other words, the system

is primed to respond rapidly. An increase in cytoplasmic Ca^{2+} signals exocytosis of the docked synaptic vessels in a process that requires a membrane protein called synaptotagmin.

15. The dendrite is the neuron extension that receives signals at synapses and the axon is the neuron extension that transmits signals to other neurons or muscle cells. Typically, each neuron has multiple dendrites that radiate out from the cell body and only one axon that extends from the cell body. At the synapse, the dendrite may have either excitatory or inhibitory receptors. Activation of these receptors results in either a small depolarization or small hyperpolarization of the plasma membrane. These depolarizations move down the dendrite to the cell body and then to the axon hillock. When the sum of the various small depolarizations and hyperpolarizations at the axon hillock reaches a threshold potential, an action potential is triggered. The action potential then moves down the axon.

16. At inhibitory synapses, the action potential in the pre-synaptic cell triggers the release of inhibitory neurotransmitters, which bind to the post-synaptic receptors activating K^+ and/or Cl^- channels. The opening of Cl^- channels results in an influx of Cl^- ions. The opening of K^+ channels results in an efflux of K^+ ions. In both cases, hyperpolarization of the post-synaptic cell occurs, making it more difficult for this cell to reach the required threshold to fire an action potential.

17. Dynamin is a GTP-binding protein that is required for pinching off of clathrin/AP-coated vesicles during endocytosis. Fly mutants that lack dynamin cannot recycle synaptic vesicles. These mutants can form clathrin-coated pits but cannot pinch off vesicles.

18.

Electrical	Chemical
Uni-directional impulse transmission	Bi-directional impulse transmission
Fast conduction velocity (0.5-5 ms)	Faster conduction velocity (fraction of a ms)
Direct electrical transmission between pre- and post-synaptic cells	Transmission between cells occurs via neurotransmitter release at the synaptic cleft, followed by the conversion of this chemical signal to an electrical signal in the post-synaptic cell
No need for the pre-synaptic cell to reach a certain threshold potential	Pre-synaptic cell must reach a threshold potential before transmission to the post-synaptic cell

19. The receptors for salt and sour taste are ion channel membrane proteins. The receptors for sweetness, bitterness, and umami taste and odor receptors are seven-membrane domain proteins. With the salt taste receptors, the direct influx

of Na^+ through the ion channel depolarizes the cell. Sour taste may work in a similar way except that depolarization is a result of H^+ flow through the receptor ion channel. The seven-membrane domain proteins are G–protein coupled receptors. They function through G proteins to release Ca^{2+} into the cytoplasm, causing depolarization of the cell.

ANALYZE THE DATA

a. TAAR6 and TAAR7 are expressed in different neurons. If this expression pattern is observed with each TAAR member, then each olfactory neuron produces only a single member of the TAAR family, just as each olfactory neuron produces only one of the classical odorant receptors. In addition, the bottom panel of images in the text figure on page 1055 suggests that the cells expressing TAAR are not the same cells that express classical odorant receptors. Thus, neurons in the nasal epithelium express a single type of olfactory receptor, be it an odorant receptor or a TAAR.

b. The data suggest that different TAARs discriminate among various amines. Cells that do not express any TAAR do not secrete alkaline phosphatase, showing that the secretion is directly linked to the presence of these receptors. Some TAARs, such as mTAAR4 and mTAAR7f, appear to be very ligand specific. mTAAR4, for example, can distinguish between two amines that differ by the presence or absence of a single hydroxyl group. Others, such as MTAAR5, which binds to more than one tertiary amine, show a broader spectrum of ligand binding. Alkaline phosphatase secretion (the SEAP assay) will occur only if alkaline phosphatase is expressed. Its expression depends on elevated cAMP levels. Such an elevation would occur if the TAARs, which are GPCRs, couple to a G protein that activates adenyl cyclase.

c. Cells expressing the mTAAR5 receptor appear to bind to a chemical present in the urine of two different species of adult male mice but not present in the urine of female mice nor in the urine of human males. This chemical may serve as a signal (pheromone) by which mice recognize the presence of sexually mature male mice. One would expect that one of the ligands that binds TAAR5, perhaps trimethylamine or methylpiperidine, would be present at appreciable concentrations in the urine of sexually mature male mice but not present in the urine of immature male mice nor in the urine of female mice nor in the urine of human males.

23

IMMUNOLOGY

REVIEW THE CONCEPTS

1. a. Pathogenic strains of *Staphylococcus aureus* secrete collagenases that can break down connective tissue, allowing entry of the bacteria.
    b. Envelope viruses have proteins that mediate fusion of the viral envelope with the host cell membrane, resulting in delivery of the viral genetic material into the host cell.

2. Like erythrocytes, leukocytes are made in the bone marrow and circulate throughout the body in the bloodstream. However, unlike erythrocytes, leukocytes can leave the bloodstream and enter lymph nodes and lymphoid organs. Here, they interact with other cells and molecules that activate immune responses. Activated cells can leave the lymph system and recirculate through the bloodstream. Pathogenic invaders evoke chemical signals that can cause functionally active leukocytes to leave the circulation, move into tissues, and attack pathogenic invaders or destroy virus-infected cells.

3. Examples of mechanical defenses include the skin, epithelia, mucus and cilia in the nose and airways, and the exoskeleton in arthropods. Chemical defenses include low pH in the stomach, lysozyme in tears, and other antimicrobial secretions.

4. The classical complement cascade is antibody dependent. Immune complexes initiate this response via recruitment of C1q. This is the first step in a cascade of proteolytic events. Recruitment of C1q allows recruitment of C1r and C1s and enables activation of their proteolytic activity. The next step involves proteolytic activation of a complex

between C2 and C4. This activation yields C3 convertase that, in turn, activates C3. Activated C3 unleashes the activities of C5–C9, ultimately resulting in the formation of a pore-forming protein complex that can attack pathogenic cells by inserting itself into biological membranes and rendering them permeable. The alternative pathway bypasses the initial steps and is not antibody dependent. This pathway starts with the spontaneous hydrolysis of the thioester bond of the C3 component of complement. An alternative C3 convertase can be formed upon interaction between the spontaneously activated C3 component and microbial surfaces. The steps downstream of C3 activation are the same as the classical pathway.

5. Emil von Behring found that serum from guinea pigs that had recovered from a diphtheria infection could convey resistance in animals that had never been exposed to diphtheria. The serum would convey protection only against diphtheria, not any other pathogen. Von Behring inferred the existence of antibodies as the factor responsible for this specific protection. If the serum was heated to >56° C, the protection was lost, but addition of fresh serum from animals not exposed to diphtheria could restore protection. From this result, von Behring concluded that there were additional heat-sensitive, nonspecific factors that work together with specific antibodies to kill pathogens. This is the complement system.

6. Decoration of particulate antigens with antibodies enhances phagocytosis. This process is called opsonization. In this process, antibodies attach to a virus or microbial surface by binding to their cognate antigen. Specialized phagocytic cells such as dendritic cells or macrophages can recognize the constant regions of bound antibodies by means of Fc receptors. Fc-receptor-dependent events allow the dendritic cells and macrophages to more readily inject and destroy antigenic particles.

7. Somatic recombination of V gene segments both completes an intact V segment and also places the promoter sequences of the rearranged V gene within controlling distance of enhancer elements required for the V gene transcription. In this way, B cells ensure that only rearranged V genes are transcribed.

8. Once a heavy-chain gene has undergone a successful recombination, it forms a complex with two surrogate light chains, λ5 and VpreB in association with Igα and Igβ. This pre-B cell receptor complex shuts off RAG expression. Since RAG expression is required for recombination, no further recombination can take place until RAG expression is reinitiated.

9. Class switch involves recombination of the gene segment for the heavy chain constant regions. The exons that encode the μ and δ heavy chains are immediately downstream of the VDJ cluster. Alternative splicing determines whether a μ or a δ chain will be produced. Downstream of the μ/δ combination are the exons that encode all of the other different isotypes. Each of these exons is preceded by repetitive sequences that promote recombination. To switch from IgM to any of the other isotypes, there is a recombination event that deletes all the intervening DNA to place the exon of any particular heavy chain constant region downstream of the VDJ cluster. (This process affects only the heavy chain.)

10. Along with other signals, T lymphocytes provide signals to antigen-activated B cells that induce expression of activation-induced deaminase (AID). AID deaminates cytosine residues to uracil. Thus, with every round of B cell replication there is a potential for mutations to accumulate. Mutations that improve the affinity of the immunoglobulin for antigen convey a selective advantage. Those B cells whose antibodies have a higher affinity for antigen tend to proliferate more. Thus, the overall affinity of a population of B cells for a particular antigen increases over time. This phenomenon is called the affinity maturation of the antibody response.

11. Both MHC Class I and MHC class II proteins are glycoproteins essential for immune recognition. Class I MHC proteins are present on all cells. In humans, Class I MHC proteins are coded by the HLA-A, HLA-B, and HLA-C loci. Class II MHC proteins are present only on antigen-presenting cells, including B cells, dendritic cells, and macrophages. In humans, class II MHC proteins are coded by six HLA-D genes.

    Cytotoxic T cells use Class I molecules as their restriction elements. T helper cells use Class II MHC molecules as their restriction elements.

12. The Class I MHC pathway presents cytosolic antigens.

    Step 1: Acquisition of antigen is synonymous with the production of proteins with errors (premature termination, misincorporation).

    Step 2: Misfolded proteins are targeted for degradation through conjugation with ubiquitin.

    Step 3: Proteolysis is carried out by the proteasome. In cells exposed to interferon γ, the catalytically active β subunits of the proteasome are replaced by interferon-induced active β subunits.

    Step 4: Peptides are delivered to the interior of the ER via the dimeric TAP peptide transporter.

    Step 5: Peptide is loaded onto newly made class I MHC molecules within the peptide-loading complex.

    Step 6: The fully assembled class I MHC–peptide complex is transported to the cell surface via the secretory pathway.

13. The Class II MHC pathway presents antigens delivered to the endocytic pathway.

    Step 1: Particulate antigens are acquired by phagocytosis and nonparticulate antigens by pinocytosis or endocytosis.

    Step 2: Exposure of antigen to the low pH and reducing environment of endosomes and lysosomes prepares the antigen for proteolysis.

    Step 3: The antigen is broken down by various proteases in endosomal and lysosomal compartments.

    Step 4: Class II MHC molecules, assembled in the ER from their subunits, are delivered to endosomal/lysosomal compartments by means of signals contained in the associated invariant (Ii) chain. This delivery targets late endosomes, lysosomes, and early endosomes, ensuring that class II MHC molecules are exposed to the products of proteolytic breakdown of antigen along the entire endocytic pathway.

Step 5: Peptide loading is accomplished with the assistance of DM, a class II MHC-like chaperone protein.

Step 6: Peptide-loaded Class II MHC molecules are displayed at the cell surface.

14. T cells that have receptors that could interact with self-MHC complexed with a particular self-peptide are identified by combining to form self-peptide MHC complexes within the thymus. If any of these combinations surpass a threshold that triggers the T-cell receptor, those cells die via an apoptotic pathway before leaving the thymus.

15. T-cell-related autoimmune diseases are associated with particular alleles of Class II MCH proteins because MHC recognition is required for T-cell attack.

16. Professional antigen-presenting cells such as dendritic cells and macrophages phagocytize pathogens and process antigens into small peptides. Interaction with pathogens activates professional antigen-presenting cells to migrate toward lymph nodes and increase the activity of their endosomal/lysosomal proteases. They also secrete cytokines that can stimulate naïve T cells. The professional antigen-presenting cells process antigens from the phagocytized pathogens into small peptides and display in the form of peptide-MHC complexes. Together with the stimulating cytokines, this sets up conditions for T cells to be activated. In the lymph nodes, B cells bind to antigens via their B-cell receptors, internalize the immune complex, and process it for presentation via the Class II MHC pathway. Activated T cells that recognize the same antigen bind to the B-cell complex, leading to B-cell differentiation and high-affinity antibody production.

17. Innate immune response:
    - Pathogen invades
    - Effector cells recognize the pathogen
    - Effector cells remove the pathogen

    Adaptive immune response:
    - Pathogen invades
    - Antigen-presenting cells identify the pathogen
    - Antigen-presenting cells travel to lymphoid organs
    - Antigen-presenting cells display antigen to lymphocytes
    - Lymphocytes proliferate and differentiate to effector cells
    - Effector lymphocytes clear pathogen

18. Passive immunization is when an antibody is administered to a person who cannot generate their own immune response, either because they do not have time (as in a snake bite toxin) or because their immune system is compromised (as in patients with an immune deficiency). Maternal antibodies passed through the placenta to a fetus are also an example of passive immunization.

19. Use noninfectious virus-like particles (virus capsid proteins devoid of any genetic material) to mimic the intact virion.

20. The vaccine is injected into the patient in order to produce a primary immune response (antibody production) that will (hopefully) neutralize the pathogen if you are exposed to it a second time. (Note: The vaccine must contain the most current strain in order to protect you; you can still get the flu after receiving a flu shot due to the many types of influenza.)

21. A polyclonal antibody can be made by injecting purified protein of interest into a mammalian model (e.g., mouse, goat, or rabbit). The mammal will produce antibodies against the protein that will be present in the blood of the animal. The blood can be drawn from that animal and purified into serum by removal of blood cells and platelets. The serum can be used directly or the antibody can be further purified.
    A monoclonal antibody can be made by injecting purified protein of interest into a mammalian model. The mammal is allowed to produce an antigen response and then the spleen is harvested. The spleen cells are dissociated and the immune B cells are fused with myeloma cells creating hybridomas, an immortal line. Each hybridoma creates a unique antibody.

22. Plasma cells synthesize and secrete antibody molecules in response to infection. It is this ramped-up production of secreted antibodies that underlies the effectiveness of the adaptive immune response in eliminating pathogens. Without plasma cells, the adaptive immune response would not be able to bind pathogen. The innate immune system targets antibody/antigen complexes for clearance, so without antibodies, the immune system would be unable to clear pathogens.

ANALYZE THE DATA

a. Because, as shown in graph C on page 1109 of the text, IL-2 is secreted in panel C and the B cells are killed, secretion of IL-2 must be from the T cells and would occur when the T cells are activated by the B cells. Activation would occur when the T cells recognize the SIINFEKL peptide, as the cells electroporated with the control protein (graph A) do not induce T-cell activation. Interestingly, the T cells are activated when SIINFEKL is presented to them on the surface of fixed cells (graph C, dashed and solid curves). Thus, B cells need not be alive, they just need to properly present antigen.

b. One of the inhibitors blocks proteolysis in lysosomes, the other blocks proteolysis induced by proteasomes in the cytoplasm. Presentation by Class I MHC involves degradation of self and foreign molecules in the cytosol by proteasomes, whereas presentation by Class II MHC involves endocytosis of microbial pathogens, which are then degraded in phagosomes/lysosomes for antigen presentation. Ovalbumin, introduced into the cytoplasm of the cells, would be expected to be cleaved by the proteasome and its peptides translocated into the ER for presentation by Class I MHC molecules. The Class I MHC-peptide complex would be transported via the secretory pathway to the cell surface. Thus, the absence of a T-cell response (graph B) in the presence of the proteasome inhibitor, but not the lysosomal inhibitor, suggests that ovalbumin follows the Class I MHC pathway, and not the Class II MHC pathway.

c. In the experiments shown in graph C, the SIINFEKL peptide, rather than intact ovalbumin, has been introduced into the B cells. Accordingly, there would be no need for the proteasome to digest ovalbumin; the appropriate peptide is already present. These data indicate that inhibition of the proteasome does not cause a deficiency in the cell that prevents peptides from being presented at the cell surface. In this case, SIINFEKL is presented regardless of proteasome function. Accordingly, the proteasome must digest the protein for antigen presentation.

24

CANCER

REVIEW THE CONCEPTS

1. Benign tumors remain localized to the tissue of origin, often maintaining normal morphology and function, and are pathological only if their sheer mass interferes with tissue function or if they overproduce a hormone or other factor that disrupts normal body homeostasis. Malignant tumors possess cells that divide more rapidly than normal, fail to die by apoptosis, invade surrounding tissues, and may metastasize to other parts of the body. The genetic difference between benign colon polyps and malignant colon carcinoma is in the number of cancer-promoting mutations. The polyp possesses a loss-of-function mutation in the *APC* gene, whereas the malignant carcinoma possesses the APC mutation as well as other cancer-promoting mutations in the *K-ras* and *p53* genes.

2. Metastasis is the process by which cancer cells escape their tissue of origin, travel through the circulation, and invade and proliferate within another tissue or organ.
   a. Batimastat inhibits enzymes that degrade the extracellular matrix, and thus cancer cells will be unable to digest the basement membrane and escape the tissue of origin.
   b. Inhibition of integrin function prevents attachment of cancer cells to the basement membrane, an early step in metastasis.
   c. Bisphosphonate inhibits osteoclasts, which are recruited and activated by many cancer cells, particularly cancers that originate in bone marrow; the cancer cells will not be able to escape from surrounding bone tissue to enter the circulation and metastasize or populate other bone marrow tissues.

d. The epithelial to mesenchymal transition (EMT) is thought to play a crucial role during the process of metastasis in certain cancers. During normal development, the conversion of epithelial cells into mesenchymal cells is a step in the formation of some organs and tissues. An EMT requires distinct changes in patterns of gene expression and results in fundamental changes in cell morphology, such as loss of cell-cell adhesion, loss of cell polarity, and the acquisition of migratory and invasive properties. During metastasis, the EMT regulatory pathways are thought to be activated at the invasive front of tumors, producing single migratory cells. At the heart of the EMT are two transcription factors, Snail and Twist. These transcription factors promote expression of genes involved in cell migration, trigger down-regulation of cell adhesion factors such as E-cadherin, and increase the production of proteases that digest the basement membrane, thus allowing its penetration by the tumor cells. Thus, EMT may be necessary but not sufficient to accelerate the metastatic process in these cancers.

3. The growth factors bFGF, TGFα, and VEGF all promote angiogenesis, the proliferation of blood vessels. If cancer cells acquire the ability to induce angiogenesis, then the tumor can develop its own vasculature and grow to a virtually unlimited size.

4. Mouse 3T3 cells normally grow in a monolayer. In a transformation assay, the 3T3 cells are transfected with DNA fragments from a human tumor. If any of the cells pick up and express a *ras* gene from the tumor DNA, they lose their contact inhibition and form a focus, a pile of cells that can be seen under the microscope. Thus, only the cells that give rise to foci contain a *ras* gene from the human tumor. In normal mouse fibroblast cells, *p53* would protect the cells against transformation. But, the 3T3 cells have loss-of function mutations in either *p53* or *p19ARF*.

5. Otto Warburg discovered that energy metabolism in cancer cells differs substantially from that in normal cells. In contrast to normal cells, most cancer cells rely on glycolysis for energy production irrespective of whether oxygen levels are high or low, producing large amounts of lactate. The metabolism of glucose to lactate generates only 2 ATP molecules per molecule of glucose, in contrast to oxidative phosphorylation, which can generate up to 36 molecules of ATP per molecule of glucose. The use of glycolysis to produce energy even in the presence of oxygen, called aerobic glycolysis, was first discovered in cancer cells by Warburg and is therefore called the "Warburg effect."

6. The increased incidence of cancer with age is explained by a "multi-hit" model; successive mutations or alterations in gene expression correspond to the discrete stages leading to a lethal tumor. For example, many colon cancers contain mutations in *APC, DCC, p53*, tumor-suppressor genes, and in *ras*. The APC mutation is found in polyps, an early stage of colon cancer, while *p53* mutation is required for malignancy. In mice, overexpression of *myc* or expression of *rasD* causes cancer only after a long lag. However, these two genes act synergistically to cause cancer in at least one-third the time of either alone.

7. Proto-oncogenes are genes that become oncogenes by mutations that render them constitutively or excessively active. They promote cell growth, inhibit cell death, or promote some other aspect of the cancer phenotype such as metastasis. Tumor-suppressing genes restrain growth, promote apoptosis, or inhibit some other aspect of the cancer phenotype. Gain of function mutations convert proto-oncogenes to oncogenes, and thus only a single copy of the proto-oncogene needs to be mutated to an oncogene to be cancer promoting. Loss-of-function mutations in tumor-suppressor genes are cancer promoting, and thus both copies of the gene usually need to be inactivated unless mutation of a single copy functions in a dominant negative manner as is the case with some mutations in the *p53* gene. The *ras, bcl-2, mdm2,* and *jun* genes are proto-oncogenes. The *p53* and *p16* genes are tumor-suppressor genes.

8. DNA microarray analysis is one method used to conduct gene expression profiling. Use of this technology enables an analysis of the relative activity levels of thousands of genes at the same time. Measuring gene expression levels in an organism provides a picture of how the organism's cells function during the cell cycle, or in response to specific stimuli. Since development of breast cancer is multi-factorial from a genetic perspective, gene expression profiling is particularly useful because women at the same stage of disease can respond very differently to treatment, and have very different prognoses as a result. Using DNA microanalysis to develop gene expression profiles has suggested that certain gene expression profile in breast cancer cells correlates with poor prognosis. Thus, application of gene expression profiling technology may help to identify women who could benefit from more aggressive treatment.

9. Cancer cells generally have lost the regulation that governs cell physiology in normal cells. This loss of regulation can occur in one or more of the following areas:
   • Transduction of growth signals to the nucleus
   • Gene expression
   • Cell cycle control
   • DNA replication and repair

   In addition, cancer cells can develop aggressive proliferative properties that differentiate them from normal cells including:
   • Ability to stimulate angiogenesis
   • EMT
   • Warburg effect

10. In hereditary retinoblastoma, individuals have inherited one mutated copy of the *RB* gene, and therefore require only a spontaneous mutation in the other copy to lack functional Rb protein. The relative frequency of a single spontaneous mutation is high enough that these individuals develop retinoblastoma early in life in both of their eyes. However, in spontaneous retinoblastoma, individuals have inherited two normal copies of the *RB* gene. Therefore, spontaneous mutations in

each copy of *RB* must occur within a single cell for it to lack functional Rb. The likelihood of a cell's possessing both mutations is extremely low, and thus these mutations rarely occur until adulthood and then usually in a single eye. Because the chance of an individual with hereditary retinoblastoma receiving an inactivating mutation in the other copy of the *RB* gene in any one of the susceptible cells is quite high, the disease is inherited in a dominant manner.

11. Many individuals are genetically predisposed to cancer because of the loss or inactivation of one copy of a tumor-suppressor gene. Loss-of-heterozygosity (LOH) describes the loss or inactivation of the second, normal copy in a somatic cell, a prerequisite for the development of a tumor because one functional copy of a tumor-suppressor gene is usually sufficient for normal function. Since the development of cancer requires loss-of-function in one or more tumor-suppressor genes (e.g., *RB*, *p53*), LOH of at least one allele is found in virtually all malignant tumors. One mechanism by which loss-of-heterozygosity develops is the missegregation of chromosomes during mitosis. The spindle assembly checkpoint normally arrests cells in mitosis until chromosomes are properly aligned on the mitotic spindle. If this checkpoint is not functional, missegregation events leading to LOH are more frequent.

12. Transmembrane growth factor receptors such as the EGF receptor are protein tyrosine kinases. Cytokine receptors such as the erythropoietin receptor activate associated JAK kinases. In the case of the EGF receptor, once activated, these receptors dimerize and activate a series of signal transduction events that ultimately result in changes in gene expression. As the EGF receptor is a tyrosine kinase, it catalyzes the phosphorylation of protein substrates once activated.

    a. The viral protein gp55 binds to the erythropoietin receptor, causing the receptor to dimerize and become activated. This leads to the constitutive activation of associated JAK kinases, even in the absence of erythropoietin. Activation of associated JAK kinases facilitates their interaction with STAT proteins, followed by migration of JAK-STAT complexes to the nucleus, where they serve to promote expression of growth-oriented genes.

    b. A point mutation in the transmembrane region of the HER2 receptor causes receptor dimerization and subsequent activation of the receptor's tyrosine kinase properties, even in the absence of EGF ligand. The result is a constitutively active HER2 receptor.

13. Gain-of-function (GOF) mutations in the *ras* gene (i.e., *rasD*), renders Ras constitutively active in the GTP-bound form. Constitutively active Ras activates the growth-promoting MAPK signaling pathway, even in the absence of upstream signals from growth factor-bound receptor tyrosine kinases. Loss-of-function mutations (LOF) in *NF1* have the same effect as GOF mutations in *ras* because *NF1* encodes a protein that hydrolyses GTP bound to Ras, converting Ras to the inactive, GDP-bound form. Since GOF mutations (such as the formation of RasD) require only a single allele to be mutated, whereas in LOF mutations (such as the inactivation of NF1) usually both alleles must be mutated, cancer-promoting mutations in *ras* are more common than cancer-promoting mutations in *NF1*.

14. The v-Src protein lacks the carboxy-terminal 18 amino acids, including tyrosine 527. Phosphorylation of tyrosine 527 on c-Src by Csk causes a conformational change that inactivates Src. Because v-Src lacks this phosphorylation site, it is insensitive to Csk and therefore constitutively active.

15. In Burkitt's lymphoma, translocation places the *c-myc* gene under the influence of the antibody heavy-chain gene-enhancers. Thus, myc is expressed at high levels, but only in cells in which antibodies are produced (e.g., B-lymphocytes). Thus, this mutation is found in lymphomas rather than in other types of cancers. myc can also be rendered oncogenic by amplification of a DNA segment containing the *myc* gene. This type of mutation is not restricted to lymphomas.

16. Smad4 is a transcription factor that transduces the signal generated when TGFβ binds to its receptor on the plasma membrane. Smad4 promotes expression of the *p15* gene, which, like *p16*, inhibits cyclin D-CDK function, promoting cell cycle arrest in G_1. Smad4 also promotes expression of extracellular matrix genes and plasminogen activator inhibitor 1 (PAI-1), both of which inhibit the metastasis of tumor cells. Thus, a loss of Smad4 abrogates both the proliferation and metastasis inhibiting effects of TGFβ signaling.

17. The INK4 inhibitor normally acts as a tumor suppressor, and loss of INK4 mimics overproduction of cyclin D1. The INK4 locus encodes at least three tumor-suppressor genes, making it the most highly vulnerable locus in the human genome. In addition to harboring the p16-encoding gene *INK4a*, immediately upstream is the INK4b locus, which encodes p15, another cyclin D–CDK4/6 inhibitor. In addition to these CDK inhibitors the locus also codes for p14ARF, a key activator of the tumor suppressor p53.

18. There are at least two ways in which viruses can transform normal cells into cancerous cells:
    1) Acutely transforming viruses—In acute cases, the virus carries an overactive oncogene, and the infected cell becomes cancerous as soon as the overactive viral gene is expressed in the cell. The oncogene may be an analogue to a wild-type human gene, like a protein involved in a signal transduction cascade. Examples may include *ras*, *jun*, or *bcl-2* genes. Once the onco-proteins are expressed, they induce activation of native cellular growth mechanisms in an unregulated fashion.
    2) Chronically transforming viruses—In these cases, the viral genome is inserted near a previously existing proto-oncogene in the genome of the infected cell. This may induce overexpression of that proto-oncogene, and subsequently uncontrolled cell division. Depending upon the exact insertion site, this event might not trigger the immediate cancerous changes. Or, if the insertion is optimally located, it might take some time for the cancerous changes to be induced. Thus, slowly transforming viruses usually cause tumors much later after infection than the acutely transforming viruses, if at all. Examples may include infection with HCV or HPV.

19. Epigenetics refers to heritable changes in gene expression that occur without alteration in DNA sequence. There are two primary and interconnected epigenetic mechanisms: DNA methylation and covalent modification of histones. Recent evidence also suggests that RNA may be intimately involved in the formation of a repressive chromatin state. Epigenetic processes may be directly involved in cancer initiation. Alternatively, changes already induced within the epigenome may "prime" cells in such a way as to promote cellular transformation upon a subsequent DNA mutagenic event. One example of how epigenetic changes may contribute to tumorigenesis is DNA methylation. Hypermethylation of CpG islands at tumor-suppressor genes switches off these genes, whereas global hypomethylation leads to genome instability and inappropriate activation of oncogenes and transposable elements.

20. The three viral proteins are E5, E6, and E7. E5 binds to PDGF receptor proteins and causes them to aggregate in the plasma membrane, which stimulates activation even in the absence of the normal growth-factor signal. E6 binds to p53 and accelerates its degradation. E7 binds to Rb and inactivates it.

21. p53 inhibits malignancy in multiple ways. When cells are exposed to ionizing radiation, p53 becomes stabilized and functions as a transcription factor to promote expression of $p21^{CIP}$—leading to cell-cycle arrest in G_1—and to repress expression of cyclin B and topoisomerase II, leading to cell-cycle arrest in G_2. Thus, p53 functions in DNA damage checkpoints during both G_1 and G_2 of the cell cycle. p53 can also promote apoptosis, in part by promoting transcription of Bax . A loss of cell-cycle checkpoints and apoptosis are both characteristics of cancer cells. The carcinogen benzo(a)pyrene is activated by enzymes in the liver to become a mutagen that coverts guanine to thymine bases, including several guanines in p53, rendering the gene nonfunctional.

22. In humans, normally only germ cells and stem cells possess telomerase activity. Telomerase maintains telomere ends and promotes immortality of cells, one characteristic of cancer. Since stem cells express telomerase, they may have a greater likelihood of becoming malignant, a concern that needs to be addressed if stem cells are to be used therapeutically to treat human disease.

ANALYZE THE DATA

1. a. Each of the three drugs causes the NSCLC cells to undergo apoptosis at a significant level. Thus, the rationale for using these drugs is that they would induce the death of the cancer cells. However, nicotine appears to interfere with this induction of apoptosis, perhaps explaining why NSCLC is resistant to chemotherapy.

   b. PARP is cleaved upon treatment with the four chemotherapeutic drugs but is not cleaved when nicotine is also present. Because PARP is cleaved upon apoptosis, these data suggest, as in (a), that the chemotherapeutic drugs induced apoptosis, but that this apoptosis does not occur if nicotine is present. p53 and p21 both appear to be produced more, or stabilized, in cells treated with the chemotherapeutic drugs. Each of these proteins would contribute to arresting the cells in G_1 or G_2, thereby keeping them from proliferating. Such an increase in the amounts of these two proteins does not occur when nicotine is also present. Thus, nicotine circumvents the ability of the chemotherapeutic drugs to induce the cells to arrest in G_1 or G_2 (which then likely leads to apoptosis). The fact that actin levels are unaffected by the various treatments indicates that the chemotherapeutic drugs specifically affect the cell cycle and nicotine interferes with this effect. Moreover, the actin data serve to ensure that the gel lanes have been loaded equally and that the differences observed in the amount of other proteins from lane to lane is biologically relevant and not a loading artifact.

c. Nicotine causes the levels of XIAP and Survivin proteins to increase (see the blot on text page 1152), and this increase is required for the survival of the cells in the presence of the chemotherapeutic drugs (the graph below shows that treatment with siRNA to prevent XIAP and Survivin increases causes cells to undergo gemcitabine-induced apoptosis even when nicotine is present). The protein-level increases appear to be mediated by the PI-3 kinase pathway because inhibition of this pathway with LY294002 eliminates the protein increases and causes apoptosis, as assessed by PARP cleavage. A likely signaling pathway is that nicotine binds to a cell surface receptor that activates PI-3 kinase which, in turn, leads to increased Survivin and XIAP, both of which help protect the cells against the apoptosis induced by chemotherapeutic drugs.

d. The patients who continue to smoke will inhale nicotine that will bind to receptors and thus activate PI-3K. PI-3K activity will result in increased Survivin and XIAP, which will oppose the action of the chemotherapeutic drugs. Normally, the drugs cause the cells to undergo apoptosis, thus destroying the tumor as well as some other cells. Thus, chemotherapy will be essentially without benefit in individuals where nicotine prevents tumor cell death.

2. a. Transgenic models can be utilized to either enhance p21 (+/+) mice or eliminate p21 (−/−) expression of certain proteins. Mutant viral proteins, including E7, can be expressed in transgenic mouse models to isolate the function of these proteins from that of normal cellular proteins. This can help to establish causal relationships between the mutations and alterations in normal cellular phenotypes, including oncogenesis.

b. This would support a loss-of-function role for p21 because this suggests that expression of p21 may have a positive impact on tumor-suppressor activity in the cell.

c. Yes this would be consistent with the hypothesis—if E7 expression did not inhibit p21, then its tumor-promoting activity would be enhanced in the p21-null background relative to the p21 (+/+) background.

d. E7 expression vectors can be constructed ex vivo and then utilized to transform cell types of interest or in animal models of cervical cancer. These expression vectors can include E7 mutations of interest.